Lecture Notes in Computer Science   12984

More information about this subseries at http://www.springer.com/series/7407

Irina Kostitsyna · Pekka Orponen (Eds.)

Unconventional Computation and Natural Computation

19th International Conference, UCNC 2021
Espoo, Finland, October 18–22, 2021
Proceedings

Springer

Editors
Irina Kostitsyna (iD)
TU Eindhoven
Eindhoven, The Netherlands

Pekka Orponen (iD)
Aalto University
Espoo, Finland

ISSN 0302-9743   ISSN 1611-3349   (electronic)
Lecture Notes in Computer Science
ISBN 978-3-030-87992-1   ISBN 978-3-030-87993-8   (eBook)
https://doi.org/10.1007/978-3-030-87993-8

LNCS Sublibrary: SL1 – Theoretical Computer Science and General Issues

This Springer imprint is published by the registered company Springer Nature Switzerland AG
The registered company address is: Gewerbestrasse 11, 6330 Cham, Switzerland

Preface

This proceedings volume contains the papers presented at the 19th International Conference on Unconventional Computation and Natural Computation (UCNC 2021) held during October 18–22, 2021. Because of the continuing COVID-19 pandemic the conference was organized as a hybrid event, with the physical meeting taking place at the Aalto University campus in Espoo, Finland, and some talks presented online over the Internet.

The UCNC conference series covers fundamental research into computation that goes beyond the standard Turing model, including both computational models and methods inspired by nature, and the computational characteristics of natural processes. The conference scope encompasses, for instance, the following topics and areas: programmable matter, material computing, molecular (DNA) computing, quantum computing, optical computing, chaos computing, collision-based computing; self-assembling and self-organizing systems, cellular automata, neural computation, evolutionary computation, swarm computing, artificial life, artificial immune systems, amorphous computing, membrane computing, physarum computing, super-Turing computation, computational neuroscience, computational systems biology, synthetic biology, and cellular (in vivo) computing.

The conference was established by Cristian S. Calude in Auckland, New Zealand, in 1998 as Unconventional Models of Computation (UMC). The second and third editions of UMC were held in Brussels (2000) and in Kobe (2002). To explicitly recognize the significance of experimental and applied work, and the inspiration from nature, the conference first changed its name to UC (Unconventional Computation) in 2005, and then to UCNC in 2012.

Since 2005 the meeting has been held annually: Seville (2005), York (2006), Kingston (2007), Vienna (2008), Ponta Delgada (2009), Tokyo (2010), Turku (2011), Orléans (2012), Milan (2013), London (2014), Auckland (2015), Manchester (2016), Fayetteville (2017), Fontainebleau (2018), and Tokyo (2019). Preparations were made for a 2020 meeting in Vienna, but due to the dramatic emergence of the COVID-19 pandemic this event had to be canceled.

Possibly due to the unusual circumstances, the submissions this year were down by about a third from previous levels, with 17 qualified contributions from authors in 10 countries: Finland, Germany, Japan, Latvia, the Netherlands, Norway, Russia, Thailand, UK, and USA. Each of the submissions was reviewed by three reviewers, and based on these and consequent discussions, the Program Committee (PC) decided to accept 12 papers for oral presentation.

In addition to the contributed presentations, the conference program included a poster session, five plenary talks by Corentin Coulais (University of Amsterdam), Cody Geary (Aarhus University), Mikko Möttönen (Aalto University), Andrew Phillips (Microsoft Research Cambridge), and Damien Querlioz (CNRS, Université Paris-Saclay). Furthermore, the conference hosted two collocated workshops:

Programmable Matter, organized by Christian Scheideler (Paderborn University) and Matthew Patitz (University of Arkansas), and the Third International Workshop on Theoretical and Experimental Material Computing (TEMC 2021), organized by Susan Stepney (University of York).

We warmly thank all authors of the contributed papers and posters, the invited speakers, the workshop organizers, and all the participants for making UCNC 2021 a pleasant and productive meeting.

As the PC co-chairs, we would also like to express our gratitude to the members of the PC and the external reviewers for reviewing the papers and participating in the selection process to help maintain the high scientific standard of the UCNC conference series. We owe a great thanks also to the EasyChair conference system, which is a wonderfully helpful tool for managing the submission and review process.

For their kind professional assistance, we would like to thank producer Laura Karvonen and controller Leila Koivisto from the Aalto University School of Science. Partial financial support was provided by the Aalto Science Institute hosted by the Aalto School of Science.

Finally, we wish to thank the editorial staff at Springer, and in particular Guido Zosimo-Landolfo and Anna Kramer, for their detailed instructions and advice in the process of publishing this volume.

August 2021

Irina Kostitsyna
Pekka Orponen

Organization

Steering Committee

Thomas Bäck	Leiden University, The Netherlands
Cristian S. Calude (Founding Chair)	University of Auckland, New Zealand
Lov K. Grover	Bell Labs, USA
Natasha Jonoska (Co-chair)	University of South Florida, USA
Jarkko Kari (Co-chair)	University of Turku, Finland
Lila Kari	University of Waterloo, Canada
Seth Lloyd	Massachusetts Institute of Technology, USA
Giancarlo Mauri	Università degli Studi di Milano-Bicocca, Italy
Gheorghe Păun	Institute of Mathematics of the Romanian Academy, Romania
Grzegorz Rozenberg (Emeritus Chair)	Leiden University, The Netherlands
Arto Salomaa	University of Turku, Finland
Tommaso Toffoli	Boston University, USA
Carme Torras	Institute of Robotics and Industrial Informatics, Spain
Jan van Leeuwen	Utrecht University, The Netherlands

Program Committee

Selim G. Akl	Queen's University, Canada
Pablo Arrighi	University of Marseille, France
Peter Banda	University of Luxembourg, Luxembourg
Daniela Besozzi	University of Milano-Bicocca, Italy
Julien Bourgeois	University of Franche-Comte, France
Olivier Bournez	École Polytechnique, France
Cristian Calude	University of Auckland, New Zealand
Matteo Cavaliere	Manchester Metropolitan University, UK
Jerome Durand-Lose	Université d'Orléans, France
Angel Goni-Morero	Technical University of Madrid, Spain
Masami Hagiya	University of Tokyo, Japan
Mika Hirvensalo	University of Turku, Finland
Natasha Jonoska	University of South Florida, USA
Jarkko Kari	University of Turku, Finland
Irina Kostitsyna (Co-chair)	TU Eindhoven, The Netherlands
Robert Legenstein	TU Graz, Austria
Makoto Naruse	University of Tokyo, Japan
Pekka Orponen (Co-chair)	Aalto University, Finland
Matthew Patitz	University of Arkansas, USA

Christian Scheideler	Paderborn University, Germany
Susan Stepney	University of York, UK
Gunnar Tufte	Norwegian University of Science and Technology, Norway

Additional Reviewers

Florent Becker
Benjamin Hellouin de Menibus
Pierre Valarcher

Abstracts of Invited Talks

Machine Materials

Corentin Coulais

University of Amsterdam, 1098 Amsterdam, XH, the Netherlands
`coulais@uva.nl`

Abstract. We will discuss how we can bridge the gap between material and machine by introducing design guidelines for materials that can be programmed to interact with their environment. We will discuss in particular how geometry, topology and analogies between mechanics and condensed matter physics can be leveraged to design metamaterials with complex and autonomous behaviors, such as programmable shape-changes, rectification, locomotion, self-oscillations and memory. We will argue that machine materials can open avenues for the design of novel energy absorption devices and of a novel generation of distributed robots.

Keywords: Mechanical metamaterials · Active matter · Topological physics · Robotic media

1 Introduction

Designing materials with advanced tasks and autonomous behavior is a tantalizing scientific and technological challenge. For instance, can a material morph into an arbitrarily complex shape? Can it locomote? Or can it store information and do basic computation? We will in this talk exploit the realm of geometry and mechanics to explore these questions. Machine materials are a versatile test bed at the table top to explore the role of geometry, topology, frustration and activity on emergent phenomena. They could also in the future provide cheaper and more sustainable materials as well as dramatically change human-matter interactions.

2 Shape-Changing and Memory

The first type of machine materials we will talk about are shape-changing materials, i.e. materials that can morph in controlled ways, see Fig. 1 for a few recent examples. The design paradigms combine mappings between the deformation of the internal degrees of freedom and spins with mechanical instabilities such as buckling. While in Fig. 1a, the metamaterial exhibits a single complex shape-change, in Fig. 1b, the metamaterial exhibits a sequence of steps and in Fig. 1c, the metamaterial can exhibit two distinct shape-changes, depending on how fast it is compressed. We will also explain how to

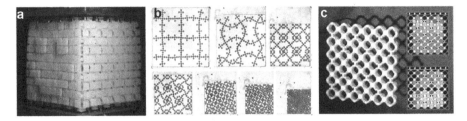

Fig. 1. Shape-changing metamaterials. (a) A cube displaying an on-demand texture upon compression. It has been designed using an analogy with spin systems [1]. (b) A metamaterial undergoing a complex sequence of foldings under compression. It exploits a combination of buckling instability and self-contact events [2]. (c) A multifunctional metamaterial that deforms in one mode upon fast compression and in another mode upon slow compression. The deformations are highlighted by red and blue ellipses overlayed on the pictures. The metamaterial has been designed using a combinatorial approach [3].

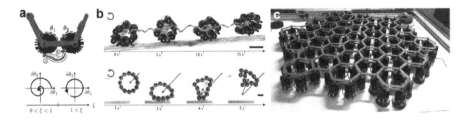

Fig. 2. Robotic media. (a) Building block of robotic media: two vertices connected by linkages and coupled asymmetrically. The building block can undergo a spontaneous limit cycle. (b) Locomotion and collision of a robotic ring. (c) Picture of a hexagonal robotic lattice. Adapted from [4].

leverage topological order and geometric frustration in similar systems to achieve programmable mechanical memory.

3   Self-oscillations and Rectification

In a second part, we will introduce the concept of robotic media, which combines the symmetry and the concept of emergence commonly used in condensed matter with the capabilities of robotics. We will discuss how nonlinear work cycles naturally emerge in such systems (Fig. 2a) and how we can use those to control how robotic media locomote and collide with a substrate (Fig. 2bc). We further discuss how such behavior is rooted in novel types of wave patterns that are rectified. We will further discuss how these concepts can be married with topology to achieve robust emerging responses.

References

1. Coulais, C., Teomy, E., de Reus, K., Shokef, Y., van Hecke, M.: Combinatorial design of textured mechanical metamaterials. Nature **535**, 529–532 (2016)
2. Coulais, C., Sabbadini, A., Vink, F., van Hecke, M.: Multi-step self-guided pathways for shape-changing metamaterials. Nature **561**, 512–515 (2018)
3. Bossartm A., Dykstram D.M.J., van der Laanm J., Coulais, C.: Oligomodal metamaterials with multifunctional mechanics. In: Proceedings of the National Academy of Sciences, USA, vol. 118, p. e2018610118 (2021)
4. Brandenbourger, M., Scheibner, C., Veenstra, J., Vitelli, V., Coulais, C.: Active impact and locomotion in robotic matter with nonlinear work cycles. arXiv:2108.08837 (2021)

Programming RNA to Fold

Cody Geary ⓘ

Interdisciplinary Nanoscience Center, Aarhus University, Denmark
geary@inano.au.dk

Living cells produce exquisitely detailed RNA nanostructures through a dynamic folding process called cotranscriptional folding. This highly efficient simultaneous assembly of many RNAs into functional structures resembles a manufacturing assemblyline, with numerous RNA transcripts being continuously produced and folded from each copy of the DNA template. Inspired by the architecture of RNAs in nature, RNA origami are a new class of RNA structure that can be rapidly designed with computers to be able fold cotranscriptionally into well defined nanostructures. Understanding cotranscriptional folding of RNA will be the key to designing functional RNA origami nanostructures that can be expressed and folded inside of living cells. Functionalized RNA nanostructures may be the next big breakthrough in nanomedicine.

In this talk I will share the latest breakthroughs in the field that have allowed us to produce highly uniform structures up to 2,360nts long. Our new software called ROAD automates the design of RNA origami sequences and also evaluates proposed structural designs for their compatibility with cotranscriptional folding. ROAD enables protein binding sites and fluorescent aptamers to be incorporated into the RNA origami structures, greatly simplifying the process of designing functional RNA nanostructures. ROAD has been validated by the production of over 40 different RNA structures that have been extensively characterized by AFM and CryoEM. Algorithmic optimization of the RNA sequence and the strand-path by ROAD appear to significantly improve the yields of correctly folded RNA origami.

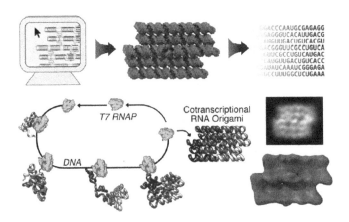

Fig. 1. The ROAD-supported RNA structure design and assembly pipeline.

Superconducting Quantum Computers

Mikko Möttönen[1,2] (iD)

[1] QCD Labs, QTF Centre of Excellence, Department of Applied Physics,
Aalto University, P.O. Box 13500, 00076 Aalto, Finland
mikko.mottonen@aalto.fi
https://www.aalto.
fi/en/department-of-applied-physics/quantum-computing-
and-devices-qcd
[2] QTF Centre of Excellence, VTT Technical Research Centre of Finland Ltd,
P.O. Box 1000, 02044, Finland

Whereas the 20th century was miraculous in introducing the humankind a new and exciting microscopic world, that of quantum physics, the 21st century will be perhaps as amazing in demonstrating the use of quantum systems, controlled at the level of single excitation quanta, in technological applications, namely, quantum technology. Quantum computing is one of the most intriguing subfields of quantum technology since it harnesses the exponentially large state space of the quantum computer to take computational shortcuts in solving otherwise difficult mathematical problems. Although several quantum algorithms that scale more favorably than their classical counterparts are known, their requirements to provide advantage in problems of practical value on a physical quantum computer still exceed the current state of the quantum hardware. Thus still major work is to be carried out in improving both, the quantum software and hardware. Here, I provide a tutorial to one of the most promising approaches to implement the hardware, i.e., superconducting quantum computers. They are built out of superconducting thin films fabricated using lithographic techniques on semiconducting or insulating chips. At millikelvin temperatures, the circuits turn superconducting and allow one to address the single excitation quanta of microwave photons trapped into the system. These photon excitations, or the lack of them, constitute the quantum bits, or qubits, that are at the heart of quantum computing. Recently, quantum supremacy has been reported in superconducting quantum computers, i.e., the quantum computer was able to solve a well-defined computational problem faster than a classical supercomputer, but this computation was not found to provide increased practical value. Nevertheless, the achievement of such a major milestone seems promising for future practical applications of superconducting quantum computers.

Keywords: Quantum computer · Superconducting circuit · cQED

Computational Design of Nucleic Acid Circuits

Andrew Phillips(iD)

Microsoft Research, Cambridge CB1 0FB, UK
andrew.phillips@microsoft.com

Information processing circuits made of nucleic acids show great potential for enabling a broad range of biotechnology applications, including smart probes for molecular biology research, in vitro assembly of complex compounds, high-precision in vitro diagnostics and, ultimately, computational therapeutics inside living cells. This tutorial presents an introduction to nucleic acid circuits and to some of the computational methods that underpin their design, simulation and analysis. The tutorial will include the following topics, which we weave together using a simple running example of a nucleic acid Join circuit:

- Theoretical underpinnings for the computational design of nucleic acid circuits and examples of circuits implemented experimentally.
- Compilation of nucleic acid circuits to chemical reaction network models.
- A comparison of simulation and analysis methods for chemical reaction network models of nucleic acid circuits: stochastic simulation, deterministic simulation, linear noise approximation and probabilistic model-checking by integration of the chemical master equation.
- Spatial simulations of nucleic acid circuits with diffusible species.
- Parameter inference and its application to nucleic acid circuit design.
- Nucleic acid circuit design abstractions: subdomains, user-defined reactions, a hierarchy of compilation abstractions, just-in time compilation, unintended reactions via leaks, circuits localized to DNA origami.
- The adaptation of logic programming to nucleic acid circuit design, by extending standard logic programming with an equational theory of strands to express nucleic acid molecular motifs.
- The use of logic programming to model circuits with complex topologies and with DNA and RNA enzymes.

Throughout the tutorial, computational methods will be illustrated with simple examples that can be executed online by attendees, and linked to practical examples of nucleic acid circuits that have been implemented experimentally. We introduce these methods using the Visual DSD system[1], which allows a broad range of computational nucleic acid circuits to be designed and analysed at the domain level. This runs in most modern browsers, allowing attendees to try out the main examples from the tutorial,

---

[1] https://classicdsd.azurewebsites.net/.

which are built into the system. Although we illustrate the methods using Visual DSD for convenience, the methods themselves are more broadly applicable. More generally, we anticipate that languages and software for programming nucleic acid circuits will accelerate the development of future biotechnology applications.

Achieving Energy-Efficient Artificial Intelligence by Exploiting Imperfect Nanodevices

Damien Querlioz ⓘ

Université Paris-Saclay, CNRS, Centre de Nanosciences et de
Nanotechnologies, Palaiseau, France
damien.querlioz@c2n.upsaclay.fr

In recent years, artificial intelligence (AI) has made tremendous progress. The capabilities of AI, however, come with a high price: a considerable energy consumption. When performing AI, computers and graphics cards consume considerably more energy for moving data between logic and memory units than for doing actual arithmetic. Brains, by contrast, achieve superior energy efficiency by fusing logic and memory entirely, performing a form of in-memory computing. Currently emerging memory nanodevices such as (mem)resistive, phase change, and magnetic memories give us an opportunity to achieve similar tight integration between logic and memory, and increase the energy efficiency of AI. However, these nanodevices come with important challenges due to their unreliable nature. In this talk, we will look at neuroscience inspiration to extract lessons on the design of in-memory computing systems with unreliable devices. We will first study the reliance of brains on approximate memory strategies, which can be reproduced for artificial intelligence. We will also highlight that brains exploit the biophysics of their components to a much larger extent than our electronic devices exploit the physics of their components [1].

We will then look at examples of systems exploiting nanotechnology in a way inspired by these insights from the brain. We will give the example of a hardware binarized neural network relying on resistive memory. Binarized neural networks are a class of deep neural networks discovered in 2016, which can achieve state-of-the-art performance with a highly reduced memory and logic footprint with regards to conventional artificial intelligence approaches. Based on measurements on a hybrid CMOS and resistive hafnium oxide memory chip exploiting a differential approach, we will see that such systems can exploit the properties of emerging memories without the need for error-correcting codes, and achieve extremely high energy efficiency [2]. Then, we will present a second approach where the probabilistic nature of emerging memories, instead of being mitigated, can be fully exploited to implement a type of probabilistic learning. We show that the inherent variability in hafnium oxide memristors can naturally implement the sampling step in the Metropolis-Hastings Markov Chain Monte Carlo algorithm, and train experimentally an array of 16,384 memristors to recognize images of cancerous tissues using this technique [3]. These results highlight the importance of understanding and embracing the unreliable nature of emerging devices in artificial intelligence designs.

References

1. Markovi, D., Mizrahi, A., Querlioz, D., Grollier, J.: Physics for neuromorphic computing. Nat. Rev. Phys. **2**, 499 (2020)
2. Hirtzlin, T., et al.: Digital biologically plausible implementation of binarized neural networks with differential hafnium oxide resistive memory arrays. Front. Neurosci. **13**, 1383 (2020)
3. Dalgaty, T., et al.: In situ learning using intrinsic memristor variability via Markov chain Monte Carlo sampling. Nat. Electron. **4**, 151 (2021)

Contents

The Complexity of Multiple Handed Self-assembly 1
  David Caballero, Timothy Gomez, Robert Schweller, and Tim Wylie

Computing with Magnetic Thin Films: Using Film Geometry
to Improve Dynamics 19
  Matthew Dale, Simon O'Keefe, Angelika Sebald, Susan Stepney,
  and Martin A. Trefzer

Robust Real-Time Computing with Chemical Reaction Networks 35
  Willem Fletcher, Titus H. Klinge, James I. Lathrop, Dawn A. Nye,
  and Matthew Rayman

Zero-Knowledge Proof Protocol for Cryptarithmetic Using Dihedral Cards. . . 51
  Raimu Isuzugawa, Daiki Miyahara, and Takaaki Mizuki

Quantum Algorithm for Dyck Language with Multiple Types of Brackets . . . 68
  Kamil Khadiev and Dmitry Kravchenko

Affine Automata Verifiers 84
  Aliya Khadieva and Abuzer Yakaryılmaz

String Assembling Systems: Comparison to Sticker Systems
and Decidability 101
  Martin Kutrib and Matthias Wendlandt

Fractal Dimension of Assemblies in the Abstract Tile Assembly Model 116
  Daniel Hader, Matthew J. Patitz, and Scott M. Summers

Bistable Latch Ising Machines 131
  Jaijeet Roychowdhury

Physical ZKP for Connected Spanning Subgraph: Applications to Bridges
Puzzle and Other Problems. 149
  Suthee Ruangwises and Toshiya Itoh

Non-instantaneous Information Transfer in Physical Reservoir Computing . . . 164
  Susan Stepney

Quantum Logical Depth and Shallowness of Streaming Data by One-Way
Quantum Finite-State Transducers (Preliminary Report)   177
  Tomoyuki Yamakami

Author Index   195

The Complexity of Multiple Handed Self-assembly

David Caballero, Timothy Gomez[(✉)], Robert Schweller, and Tim Wylie

University of Texas Rio Grande Valley, Edinburg, TX 78539-2999, USA
{david.caballero01,timothy.gomez01,robert.schweller,
timothy.wylie}@utrgv.edu

Abstract. In this paper we study complexities for the multiple-handed tile self-assembly model, a generalization of the two-handed tile assembly model in which assembly proceeds by repeatedly combining up to h assemblies together into larger assemblies. We first show that there exist shapes that are self-assembled with provably lower tile type complexities given more hands: we construct a class of shapes S_k that requires $\Omega(\frac{k}{h})$ tile types to self-assemble with h or fewer hands, and yet is self-assembled in $\mathcal{O}(1)$ tile types with k hands. We further examine the complexity of self-assembling the classic benchmark $n \times n$ square shape, and show how this is self-assembled in $\mathcal{O}(1)$ tile types with $\mathcal{O}(n)$ hands. We next explore the complexity of established *verification* problems. We show the problem of determining if a given assembly is *produced* by an h-handed system is polynomial time solvable, whereas the problem of *unique assembly verification* is coNP-complete if the hand parameter h is encoded in unary, and coNEXP-complete if h is encoded in binary.

1 Introduction

In this paper we investigate the complexity of fundamental problems related to multiple handed self-assembly. The model is a tile self-assembly model where system components are 4-sided Wang tiles, and self-assembly proceeds by tiles combining non-deterministically, based on matching glue types, to build larger assemblies. The most studied tile assembly models are the aTAM [26] where tiles attach one-by-one to a growing seed assembly, and the 2HAM [6] where assemblies are produced by taking any two assemblies (one in each hand) and combining them to create a new producible assembly. Our focus here is a generalization of the 2HAM, called the k-HAM, which allows groups of up to k assemblies (one in each of up to k hands) to combine to create new stable assemblies. See [20] for a general survey of tile self-assembly, and [27] for a survey of intrinsic simulation in tile self-assembly, and [11,28] for an overview of algorithmic self-assembly and recent experimental implementations in DNA.

This research was supported in part by National Science Foundation Grant CCF-1817602.

I. Kostitsyna and P. Orponen (Eds.): UCNC 2021, LNCS 12984, pp. 1–18, 2021.
https://doi.org/10.1007/978-3-030-87993-8_1

Why Multiple Hands? The most fundamental version of self-assembly may be the basic 2-handed model as the statistical likelihood of multiple pieces combining within a short enough time to stabilize is small in many experimental models. However, such productions happen in practice- often referred to as undesirable *spurious nucleation reactions* [19]. At certain scales, it is useful to consider the landscape of *stable* configurations, especially *local minimum energy* configurations, even when reaching such configurations would require moving more than two things into place simultaneously. Multiple handed self-assembly provides a simple framework for exploring self-assembly phenomena that may have impacts in these scenarios. Notably, multiple handed self-assembly may prove to be a useful tool for designing robustness in self-assembly systems, e.g., designing a system guaranteed to work even if the number of hands is increased up to some value k. Applications of such robustness theory would be directly aided by understanding fundamental complexities of self-assembly with multiple hands.

There are also classes of shapes that may not be built efficiently without multiple hands such as certain fractals that rely on multiple shapes coming together simultaneously [8], and there are shapes that cannot be built as efficiently at lower temperatures. Thus, given different experimental constraints, a reliance on these interactions with fewer tiles at a lower temperature may be preferable.

Our Results. We provide many results that explore multiple handed assembly from two angles: the complexity of fundamental problems, and complexity separation between the 2HAM and the h-HAM related to certain classes of shapes. Results are shown compared to previous work in Tables 1 and 2 respectively.

The first set of results within this model involve exploring the complexity of the computational problems of *producibility*, asking whether a given system of tiles produces a given assembly, and the *unique assembly verification* (UAV) problem asking whether a given tile system uniquely produces a given assembly (i.e. all producible assemblies can continue to grow into the single provided target assembly). We show that the producibility problem is solvable in polynomial time, and that the UAV problem is coNP-complete if the parameter k (number of hands) is encoded in unary, and coNEXP-complete if k is encoded in binary. In particular, these hardness results hold for the standard 2D scenario with a $\mathcal{O}(1)$-bounded temperature parameter. In comparison, while the UAV problem is known to be coNP-complete for both the 3D 2HAM [6] and the 2D 2HAM for a non-constant bounded temperature [22], the complexity of UAV in the 2HAM for 2D $\mathcal{O}(1)$-bounded temperature is still open.

Our next set of results show that there exist shapes that can be built more efficiently with more hands. We first provide a class of shapes, S_k, such that S_k requires at least $\Omega(\frac{k}{h})$ unique tile types to be assembled for any system with at most h hands, and yet is buildable in $\mathcal{O}(1)$ tile types with a k-handed system. Next we consider the classic benchmark of an $n \times n$ square, and show this shape can be built with $\mathcal{O}(1)$ tile types and $\mathcal{O}(n)$-hands, which is provably fewer tile types than needed with a $\mathcal{O}(1)$-handed system for almost all integers n [21].

Related Work. The h-handed self-assembly model was introduced in [8] as a generalization of the 2HAM, and was used to build a version of the Sierpinski

Table 1. Tile complexity for $n \times n$ squares.

Model	Tile complexity	Thm.
aTAM/2HAM	$\Theta(\frac{\log n}{\log \log n})$	[21]
2HAM (high temperature)	$\mathcal{O}(2^{\log^* n})$	[23]
kHAM	$\Theta(1)$	Theorem 2

Table 2. Related and new results for verification problems. *Two results show coNP-hardness of UAV in the 2HAM, one uses a step into the third dimension and the other uses the temperature as part of the input.

Problem	Hands	Results	Thm.
Producibility	aTAM	P	[1]
Producibility	2	P	[12]
Producibility	k	P	Theorem 3
UAV	aTAM	P	[1]
UAV	2	coNP-Complete*	[6,22]
UAV	k (Unary)	coNP-Complete	Corollary 2
UAV	k (Binary)	NEXPTIME-Complete	Theorem 4

Triangle fractal. Earlier work compared 2-hands (2HAM) over a single hand (aTAM) and showed a provable gap in tile complexity for building certain shapes [6], coNP-completeness for the UAV problem in the 3D 2HAM with a constant temperature [6] and coNP-completeness in 2D for non-constant temperature [22], versus a polynomial time solution to UAV in the aTAM [1]. For the case of the *producibility* problem, both the 2HAM and aTAM have polynomial time solutions [1,12]. Later work considered *Unique Shape Verification*, showing coNP^{NP}-completeness for the 2HAM [24] and coNP-completeness for the aTAM [2].

In [5], the authors show a separation in the number of tile types needed to construct some shape between the deterministic aTAM (only one terminal assembly) and the non-deterministic version of the model (allows for multiple terminal assemblies all with the same shape).

Building infinite classes of shapes with a fixed size-$\mathcal{O}(1)$ set of tile types has been explored in several self-assembly models. For example, [15,25] show how a fixed tile set can be programmed to build general shapes by adjusting the temperature of the system over a sequence of stages. Similarly, [9] builds general shapes with a fixed set of tiles by mixing combinations of the tile set into different bins over a sequence of stages. In [7], arbitrarily large squares are self-assembled with a fixed tile set size by encoding the desired square width into the system temperature. In [14], large shapes are self-assembled with a fixed tile set by considering bonding functions between assemblies that require greater strength to hold together larger assemblies. In [4,10,16,18], fixed tile sets are used to build general squares and shapes with high probability by encoding the desired target shape into the relative concentrations of the tiles within the system.

2   Definitions

Tiles. A *tile* is a non-rotatable unit square with each edge labeled with a *glue* from a set Σ. Each pair of glues $g_1, g_2 \in \Sigma$ has a non-negative integer *strength* $\mathrm{str}(g_1, g_2)$.

Configurations, Bond Graphs, and Stability. A *configuration* is a partial function $A : \mathbb{Z}^2 \to T$ for some set of tiles T, i.e. an arrangement of tiles on a square grid. For a given configuration A, define the *bond graph* G_A to be the weighted grid graph in which each element of $\mathrm{dom}(A)$ is a vertex, and the weight of the edge between a pair of tiles is equal to the strength of the coincident glue pair. A configuration is said to be τ-*stable* for positive integer τ if every edge cut of G_A has strength at least τ, and is τ-*unstable* otherwise.

Assemblies. For a configuration A and vector $\vec{u} = \langle u_x, u_y \rangle$ with $u_x, u_y \in \mathbb{Z}^2$, $A + \vec{u}$ denotes the configuration $A \circ f$, where $f(x,y) = (x + u_x, y + u_y)$. For two configurations A and B, B is a *translation* of A, written $B \simeq A$, provided that $B = A + \vec{u}$ for some vector $\vec{u}$. For a configuration A, the *assembly* of A is the set $\tilde{A} = \{B : B \simeq A\}$. An assembly $\tilde{A}$ is a *subassembly* of an assembly $\tilde{B}$, denoted $\tilde{A} \sqsubseteq \tilde{B}$, provided that there exists an $A \in \tilde{A}$ and $B \in \tilde{B}$ such that $A \subseteq B$. An assembly is τ-*stable* provided the configurations it contains are τ-stable. Assemblies $\tilde{A}$ and $\tilde{B}$ are τ-*combinable* into an assembly $\tilde{C}$ provided there exist $A \in \tilde{A}$, $B \in \tilde{B}$, and $C \in \tilde{C}$ such that $A \cup B = C$, $A \cap B = \varnothing$, and $\tilde{C}$ is τ-stable. Let the shape of an assembly A be the shape taken from the set of points in $dom(A)$.

Two-Handed Assembly. A Two-handed assembly system $\Gamma = (T, \tau)$ is an ordered tuple where T is the *tile set* and τ is a positive integer parameter called the *temperature*. For a system Γ, the set of *producible* assemblies P'_Γ is defined recursively as: 1) $S \subseteq P'_\Gamma$. 2) If $A, B \in P'_\Gamma$ are τ-combinable into C, then $C \in P'_\Gamma$.

k-Handed Assembly. The k-handed assembly model is a generalization of two-handed assembly model. A k-handed assembly system $\Gamma' = (T, k, \tau)$ is an ordered tuple where T is the *tile set*, k is the number of hands that can be used to produce an assembly, and τ is a positive integer parameter called the *temperature*. For a system Γ', the set of *producible* assemblies $P'_{\Gamma'}$ is defined recursively as follows:

1. $S \subseteq P'_{\Gamma'}$.
2. For $2 \leq k' \leq k$, if $\{A_1, A_2, \ldots, A_{k'}\} \subset P'_{\Gamma'}$ are τ-combinable into C, then $C \in P'_{\Gamma'}$.

A producible assembly is *terminal* provided it is not τ-combinable with any other producible assembly, and the set of all terminal assemblies of a system

Γ is denoted P_Γ. Intuitively, P'_Γ represents the set of all possible assemblies that can self-assemble from the initial set T, whereas P_Γ represents only the set of assemblies that cannot grow any further. The assemblies in P_Γ are *uniquely produced* iff for each $x \in P'_\Gamma$ there exists a corresponding $y \in P_\Gamma$ such that $x \sqsubseteq y$. Thus unique production implies that every producible assembly can be repeatedly combined with others to form an assembly in P_Γ.

Unique Production of Shapes and Assemblies. A system Γ uniquely assembles an assembly A if the system uniquely produces set P_Γ that contains only the assembly A and no other assemblies. In other words all producible assemblies can be combined to eventually form A. We say a system uniquely assembles a shape S if the system uniquely produces set P_Γ and for all $B \in P_\Gamma$, B has the shape S.

Fig. 1. Shape $\mathcal{S}_k$ is constructed by connecting width-3 loops of decreasing height starting at $\frac{k-6}{2}$. The base shape is highlighted by a red dotted box. Loops are shown in light gray. The darker column of 3 tiles on the left row is the cap column. The rest of the tiles are used to connect the loops. (Color figure online)

3   Shape Building

In this section we show a separation between systems with a differing number of hands. We start by defining a shape $\mathcal{S}_k$ and then proving the lower bound on the number of tiles needed to construct the shape in relation to the number of hands used, which is used to prove the separation.

3.1   Separation

We define the shape $\mathcal{S}_k$ for all even numbers $k \geq 14$ and for the smallest shape with $k = 11$. The shape for a given k, $\mathcal{S}_k$, is described in Fig. 1, and is built recursively in a $\tau = 3$ system. The smallest shape, $\mathcal{S}_{11}$, is highlighted in the figure and is a 3×3 loop with an additional 3 tiles on its left side which we will call the cap column. $\mathcal{S}_k$ is constructed by adding an additional height $\frac{k-6}{2}$ height loop on the left side of $\mathcal{S}_{k-1}$ and connecting it with 3 tiles (darker tiles in figure). Let $min_h(\mathcal{S}_k)$ be the minimum number of tile types needed to uniquely construct an assembly of shape $\mathcal{S}_k$ in an h-handed system.

Lemma 1. *For any h-handed system $\Gamma = (T, h, 3)$ that uniquely assembles the shape $\mathcal{S}_k$, $|T| \geq \Omega(\frac{k}{h})$*

Proof. Since Γ uniquely assembles the shape $\mathcal{S}_k$, each assembly in the set of terminals P'_Γ are of that shape. Consider the rightmost column of any assembly $A \in P'_\Gamma$, which we will call c, and let g be the number of strength $< \tau$ glues between tiles in this column. c can be divided into $g + 1$ segments that are connected using strength τ glues. Each segment is a producible assembly.

Since each of the segments have a strength τ (or greater) glue in between each other they cannot attach to the other segments unless a loop is formed. Let B be any producible assembly such that B is the subassembly of the shape $\mathcal{S}_k$, but does not contain c. Since our system has h hands we are able to attach up to $h - 1$ segments to B in a single production step. Since the total length of the column is $\frac{k-6}{2}$, there must be a segment of length $\geq \frac{k-6}{2(h-1)}$. In order to build this segment, there must not be any repeated glues within that segment, otherwise the system could produce an infinitely growing assembly. Therefore, the number of tiles needed to construct this assembly is $\Omega(\frac{k}{h})$. □

3.2   Upper Bound for Building $\mathcal{S}_k$

The tile set $T_\mathcal{S}$ is shown in Fig. 2a. Let the assembly A_k be an assembly of shape $\mathcal{S}_k$ shown in Fig. 2b.

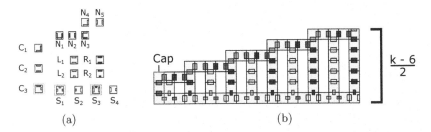

(a)   (b)

Fig. 2. (a) Constant sized tile set to construct an assembly with shape $\mathcal{S}_k$ with k hands. Larger rectangles represent glues of strength 2, while smaller rectangle represent strength 1. (b) Assembly of shape $\mathcal{S}_k$ made from the tile set.

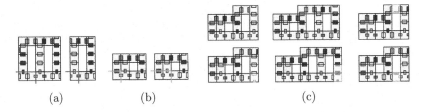

(a)   (b)   (c)

Fig. 3. (a) If one of these tiles are the bottom corner there will be a cut of strength 2 making the assembly not stable. (b) If the cap is on the assembly there does not exist a cut and the assembly is stable. (c) These are the possible conflicting tiles when attempting to construct a rogue assembly. The red line in each of these assemblies separates the column c from the rest of the assembly. (Color figure online)

Lemma 2. *Any τ-stable assembly in the h-handed system $\Gamma = (T_S, h, 3)$ is a subassembly of A_x for some $x > h$ and must contain the Cap column.*

Proof. Due to space constraints, we omit the proof and give the notes: 1) $\tau = 3$ and every glue is strength 1 or 2, so any stable assembly must contain a loop since any tile connected at only one point is not stable. 2) Only three tiles can be the bottom left corner of the loop (Fig. 2a). 3) There is a cut for all tiles unless the cap tile is present (Figs. 3a, 3b, and 3c). □

Lemma 3. *For all even $k \geq 12$ there exists a k-handed self-assembly system $\Gamma = (T, k, 3)$ uniquely assembling an assembly of shape $\mathcal{S}_k$ using $\mathcal{O}(1)$ tile types.*

Proof. Due to space constraints, this proof is not given, but we provide the tile set T in Fig. 2a as part of the system $\Gamma = (T, k, 3)$ that uniquely constructs the assembly seen in Fig. 2b using k hands. □

Theorem 1. *For all even $k \geq 12$ and $h < k$, there exists a shape $\mathcal{S}_k$ such that $min_h(\mathcal{S}_k) = \Omega(\frac{k}{h})$ and $min_k(\mathcal{S}_k) = \mathcal{O}(1)$. For the special case of $h = 2$, $min_2(\mathcal{S}_k) = \Omega(k)$.*

Proof. From Lemma 1, in the 2HAM the lower bound for constructing an assembly of shape $\mathcal{S}_k$ is $\Omega(k)$. From Lemma 3, the upper bound for uniquely constructing the shape is $\mathcal{O}(1)$ (Fig. 4). □

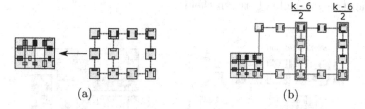

Fig. 4. (a) Using 11 hands, the base case of the assembly is built from single tiles. Using this as a single assembly, the next loop can be built. (b) For all future loops, they must be built by taking the previous sized assembly, the 5 tiles used to connect the two columns, and enough column tiles to connect them. This means that 6 hands are used to attach non-column tiles/assemblies and the remaining hands are used to build the two columns resulting in a max height of $\frac{k-6}{2}$.

3.3 Building Squares

In this section, we show that there exists a constant-sized tile set that can uniquely assemble the shape of an $n \times n$ square, where n is based on the given parameter specifying the number of hands of the system.

Theorem 2. *There exists a tile set T, consisting of 72 tile types, such that for all even integers $n \geq 10$, the h-handed tile assembly system $\Gamma = (T, h = n+1, \tau = 3)$ uniquely assembles an assembly A that has the shape of an $n \times n$ square.*

Proof. We prove by construction giving the tile set T (Fig. 5a). Solid lines represent unique strength-3 glues between the tiles. The tile set and final assembly consist of three sections. The base assembly is a 6×6 square that connects the other two sections. Both the horizontal and vertical sections build a staircase shaped structure, similar to Sect. 3.1, where each "step" of the staircase consists of a loop of tiles, and the largest buildable step is determined by the number of hands. This construction does not have space in the loops though, which creates rectangles increasing in size. We scale the size of the step by 2 and the vertical section 2 tiles taller in order for the three sections to fit together.

The tiles in the horizontal section build a staircase shaped assembly with increasing height. Attaching to the right, the tiles in the vertical section build a rotated staircase shaped assembly of increasing width as it builds upwards. This process continues until the addition of the next step requires more hands than allowed (Fig. 6). The two staircase assemblies and base assembly fit together to form a square shaped terminal assembly. An example is shown in Fig. 5b.

It is easily verifiable that in an h-handed system, an $(h-1) \times (h-1)$ square is producible. The two staircase assemblies are built up from the base assembly as shown in Fig. 6. The largest step of the horizontal assembly (blue) will be $h-3$, while the largest step of the vertical assembly (red) will be width $h-1$.

The argument that this tile system *uniquely* produces A is similar to that of Theorem 3. We focus on the horizontal section (blue) since the vertical section functions identically. In this case, the placement of tiles is even more restricted as the placement of the two repeating dominoes require each other to be stable due to the strength-1 glue between them.

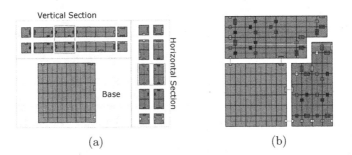

(a)   (b)

Fig. 5. (a) The tile set T that uniquely assembles an $(h-1) \times (h-1)$ square in an h-handed system. Solid lines between tiles represent a unique strength-2 glue between them. Small colored labels represent strength-1 glues, and large colored labels represent strength-2 glues. The glues in the vertical section are represented with the same colors as the horizontal section, but are hatched to signify they are distinct from their unhatched counterpart. The tiles boxed in red represent the section of each loop which can be repeated to make a loop of an arbitrary size. (b) This assembly is uniquely built in an 11-handed system. (Color figure online)

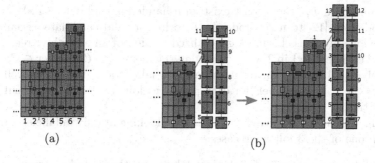

Fig. 6. (a) Cuts showing the assembly is unstable if column 1 or 3 are not the leftmost column in the horizontal section. The same cuts could larger in the assembly if there were no tiles to their left. (b) Construction of the largest horizontal staircase step in a 13-handed system. Note the largest step in the vertical section would be 12 tiles wide. (Color figure online)

A stable assembly must include the smallest step of the staircase. Figure 6a shows an assembly built from the tiles in the horizontal section. If column 2 or 5 were the leftmost column of this assembly, the red and cyan cuts show the resulting assembly is not stable, respectively. It is inherent that a stable assembly built from these tiles is either attached to the base assembly, or to column 3.

In a similar argument to the previous construction, only a few conflicting tiles exist that any rogue assembly may contain with the repeating dominoes, since without them it is clearly a subassembly of A. By starting from a pair of adjacent repeating dominoes, we work upwards and downwards, noting the possible tiles that *could* be placed, and see that they must exist in the "loop" composed of all the tiles of the horizontal section in order to be in a stable assembly.   □

4 h-Hand Producibility

Here we show that verifying producibility of an assembly is solvable in the h-handed model in polynomial time. The proof is a modification of the proof of 2-handed producibility in [12] generalized to h-handed assembly. A partition of a configuration C is a set of unique configurations $\mathcal{C} = \{C_1, C_2, \ldots, C_n\}$ such that $\bigcup_{i=1}^{h} C_i = A$ and for all $i \neq j$, $C_i \cap C_j = \emptyset$. With regards to a partition of an assembly A, we mean the partition of an arbitrary configuration $C \in A$.

Definition 1 (h-handed Assembly Tree). *An h-Handed Assembly Tree for a configuration C is a tree Υ where the root represents C, every other node represents a configuration $c \subseteq C$, every parent node has at most h children, and every parent node p has the characteristic that it's children are τ-combinable in an h-handed assembly step into p.*

Lemma 4. *For any h-handed system $\Gamma = (T, h, \tau)$ and partition C of assembly $A \in P'_\Gamma$, if $\forall a \in C, a \in P'_\Gamma$ then there exists a subset s of C such that $2 \leq |s| \leq h$ and the elements of s are τ-combinable into an assembly $B \sqsubseteq A$.*

Proof. Since $A \in P'_\Gamma$, there must exist an h-handed assembly tree Υ. We utilize a method from [12] to mark nodes in Υ to find a valid candidate assembly B. The previous proof used a more generalized version of an assembly tree.

Label each leaf $\{x\}$ in Υ with the unique element $\in C_i \in \mathcal{C}$ where $x \in C_i$. Then iteratively, if all siblings have the same label, label the parent C_i as well. This preserves the partition labels for each parent as long as it is a proper subset of the partition.

Doing a breadth first search from the root looking at only unlabeled vertices, we reach one of the 3 following cases,

1. The children have $2 \le b \le h$ distinct labels and there are b children.
2. The children have $2 \le b \le h-1$ distinct labels and there are $3 \le c \le h$ children and $c > b$.
3. There are labeled and unlabeled children or all unlabeled children. We ignore these nodes since there must exist nodes from either Case 1 or 2 if we follow the unlabeled children since all leaves are labeled.

Case 1. This case shows that there must exist b partitions that neighbor each other in the tree and can be brought together with b hands. There is a subtle subcase that for the parent assembly node p, some number of the children could be combined with fewer hands. However, a modification to the build path in this way does not change p since it could be built from the single b-handed operation or through multiple joins of less than b hands since each operation is joining subsets of different partitions and p can still be formed as a stable assembly.

Case 2. Even though $3 \le c \le k$ hands are needed, some of the nodes have the same label. Thus, the number of distinct partition subsets is $2 \le b \le h-1$. Similar to Case 1, some of the children could be combined without assembling all b children at once. Any stable combination of children represents another valid h-handed assembly sub-tree for the parent node.

Let s' be the set of configurations represented by the children of the found node. For each element in s' replace it with the element of C it was labeled by (only once for each label) to form the set s. This replacement preserves the ability for all the assemblies in s to be combinable. Since we know each element

---

Result: Given an h-handed assembly system $\Gamma = (T, h, \tau)$, and an assembly A, is A producible by Γ?

```
/* Subassemblies of A as positions. Initially individual tiles.  */
C ← {{v}|v ∈ dom(A)};
while |C| > 1 do
    if ∃ 2 ≤ b ≤ h subassemblies in C (denoted Cᵢ ∈ C with 1 ≤ i ≤ b) s. t.
    ∪₁≤ᵢ≤ᵦCᵢ is stable then
        |  C ← C \ {C₁,...,Cᵦ} ∪ {∪₁≤ᵢ≤ᵦCᵢ}
    else reject
accept;
```

Algorithm 1: The naïve method of verifying whether an assembly is producible in an h-handed system.

of s are producible, the assembly B is producible where the elements of s are τ-combinable into B. $\square$

Theorem 3. *The producibility problem for a system $\Gamma = (T, h, \tau)$ and assembly A is solvable in $\mathcal{O}(|A|^2 h \log |A|)$ time.*

Proof. Algorithm 1 gives the naïve method for building the shape by combining tiles from the shape whenever possible. We know from Lemma 4 that if the target assembly A is producible, there must exist up to h subassemblies that may be combined at each step.

The runtime is affected by the time required to find cycles in planar graphs. In order for assemblies to be combined they must be adjacent. Any assembly step that requires more than 2 hands must form a loop. Thus, the bottleneck is checking for a cycle of size h in a planar graph, which varies based on h. We must also check for cycles up to size h, so we might require h calls to this subroutine. Let T be the time to find a cycle, then the runtime of Algorithm 1 is $\mathcal{O}(Th|A|)$.

Arbitrary fixed length cycles in planar graphs with n nodes can be found in $\mathcal{O}(n \log n)$ time (with expected $\mathcal{O}(n)$ time), and for any $h \leq 6$, the complexity is $\mathcal{O}(n)$ [3,17]. Thus, for an unknown h, the runtime of Algorithm 1 is $\mathcal{O}(|A|^2 h \log |A|)$ as the size of the graph is the size of the assembly.

We note that there is a $\mathcal{O}(n)$ algorithm to find any fixed subgraph H in a planar graph, but it requires an extremely large constant that is generally considered impractical even for small n [13]. Also in the special case of 2-handed assembly the runtime of the algorithm shown in [12] runs in time $\mathcal{O}(n \log^2 n)$. $\square$

5 h-Hand Unique Assembly Verification

In this section we investigate the complexity of the problem of verifying an assembly is uniquely assembled by a given h-HAM system. We consider two different methods of encoding the number of hands in the system. We show that the problem is coNEXP-complete and coNP-complete when the number of hands is encoded in binary and unary, respectively. The problems are listed below. We first show membership, then prove hardness with a reduction from K-A$_{NTM}$.

*Problem 1 (*H-HAM-UAV*).* **Input:** An h-HAM system $\Gamma = (T, h, \tau)$ where the integer h is encoded in binary, and an assembly A. **Output:** Does Γ unique produce the assembly A?

*Problem 2 (*U-H-HAM-UAV*).* **Input:** An h-HAM system $\Gamma = (T, h, \tau)$, where the integer h is encoded in unary, and an assembly A. **Output:** Does Γ unique produce the assembly A?

5.1   Membership

The UAV problem is in the class coNEXP if the number of hands is specified in binary, and in coNP if is specified in unary. For an instance of UAV ($\Gamma =$

$(T, h, \tau), A)$, the instance is true if and only if the following 3 conditions are true: 1) The target assembly A is producible, 2) there does not exist a terminal assembly $C \sqsubset A$, and 3) there does not exist a producible assembly $B \not\sqsubseteq A$.

Lemma 5. H-HAM-UAV $\in$ *coNEXP*.

Proof. We provide a coNEXP algorithm for H-HAM-UAV that checks the above three conditions. By Theorem 3, condition 1 can be decided in polynomial time. Utilizing Lemma 4, we show that if condition 1 is true then condition 2 is true. For some assembly $C \sqsubset A$, consider any partition of the assembly A where C is an element, and by repeatedly applying Lemma 4, continue joining elements of this partition until A is built. To do this, C must, at some point, attach to another assembly. Therefore, C is not terminal.

The remaining task to decide the instance of UAV is to verify that there does not exist a producible assembly $B \not\sqsubseteq A$. A coNEXP machine can do this by nondeterministically attempting to build an assembly up to size $h|A|$. If any branch builds some assembly $B \not\sqsubseteq A$, then the branch (and machine) will reject. It suffices to check only up to this size, as an assembly of size $> h|A|$ must have been built from at least one assembly of size $> |A|$. That assembly itself is not a subassembly of A, and therefore if it exists, then a different branch of the computation will build it and reject. Since h is encoded in binary it takes exponential time to build an assembly of size $h|A|$. $\square$

Corollary 1. U-H-HAM-UAV $\in$ *coNP*.

Proof. The proof of this is the same algorithm provided in Lemma 5. Since the integer h is encoded in unary, nondeterministically building an assembly of up to size $h|A|$ takes polynomial time. $\square$

5.2   Hardness: Reduction from k-A_{NTM}

To show coNEXP-hardness we reduce from the canonical complete problem for NEXP, K-A_{NTM}, which is the problem of deciding if there exists a computation path of length $\leq k$ where a given nondeterministic Turing machine M accepts when run on the empty tape. We first overview the construction, and then prove correctness. The formal problem definitions follow.

*Problem 3 (*K-A_{NTM}*).* **Input:** A nondeterministic Turing machine M, and an integer k encoded in binary. **Output:** Does there exist a computation path of M accepting within k steps when run on an empty tape?

*Problem 4 (*U-K-A_{NTM}*).* **Input:** A nondeterministic Turing machine M, and an integer k encoded in unary. **Output:** Does there exist a computation path of M that accepts within k steps when run on an empty tape?

Given an instance of K-A_{NTM}, we create an instance of H-HAM-UAV such that the system is temperature-4, and the number of hands h is set to $(2^{\lceil \log_2(k) \rceil} + 3) \cdot (k + \lceil \log_2(k) \rceil + 4)$. The system always builds a specific target assembly. If and only if the answer to K-A_{NTM} is yes, the system also produces a *computation assembly*, an assembly that represents an accepting computation path of M that is less than k steps. The computation assembly is *not* the target assembly, making the answer to the UAV instance 'no.' We will walk through an example, reducing from an instance $(M, k = 4)$ and creating a temperature-4 system where the number of hands is set to $(4 + 3) \cdot (4 + 2 + 4) = 70$.

We first explain the case when the instance of K-A_{NTM} is true, and therefore the created instance of H-HAM-UAV is false. In this case, a computation assembly is built. A computation assembly is composed of a binary counter section and a Turing machine simulation section (Fig. 7c). Since there exists an accepting computation path of less than or equal to 4 steps, then in one production step, utilizing the large number of hands in the system, ≤ 70 tiles will come together forming a tableau that represents a simulation of the computation path, as well as a binary counter enforcing the simulation to maintain a certain tape width.

Binary Counter. We utilize known techniques, such as in [21], for implementing a self-assembling binary counter where the construction has a size-$\mathcal{O}(\log_2 k)$ tile set and bounds the counter such that it stops once it reaches $2^{\lceil \log_2(k) \rceil}$. For our example instance $(M, k = 4)$, the assembled binary counter is shown in Fig. 7a. This assembly is one of two parts of the computation assembly, and is not stable by itself at temperature 4. The binary counter counts from left to right, starting at 0 and ending at $2^{\lceil \log_2(k) \rceil}$. The bottom row of light gray tiles represents the least significant bit, while the top row represents the most significant bit. Each row uses a distinct set of tiles, preventing unbounded growth. The majority of the tiles (light gray) have a strength-1 glue on each side. Thus, these tiles are adjacent to a tile on every side in order to be stable in the assembly. The remaining border tiles (dark gray) are the only tiles that can be on the border of a stable assembly due to their strength-2 glues. Every tile in the top row is not connected to the rest with the required strength of 4. As depicted, the assembly can only be stable in combination with an additional assembly above it.

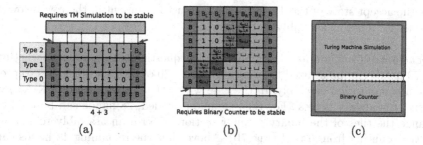

Fig. 7. (a) Example binary counter that counts up to 4. Each type consists of its own $\mathcal{O}(1)$-sized set of tiles. Single ticks between tiles represent a strength-1 glue, double ticks represent strength-2 glues. (b) Example Turing machine simulation. (c) The form of a computation assembly. (Color figure online)

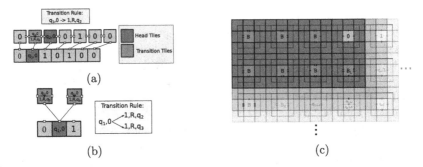

Fig. 8. (a) One step in the simulation of a Turing machine. The bottom row represents an initial valid Turing machine configuration. The tiles that can attach above this row represent a valid transition to another valid configuration of the Turing machine. (b) Simulation of nondeterministic transition rules. The glues on the south side of both of the transition tiles (purple) are the same, allowing either to be placed above the head tile (green). (c) Example target assembly. Solid lines represent unique strength-4 glues. Every tile type used in the TM simulation section and binary counter section can attach in only one spot. (Color figure online)

Turing Machine Simulation. We also use known techniques for simulating Turing machines in a self-assembly system [26]. An example of simulating one step is shown in Fig. 8a. We use this method to simulate the computation paths of M. Due to the nondeterministic nature of the model, we simulate nondeterministic transition rules by simply having a different tile type for each possible transition (Fig. 8b). For the instance $(M, k = 4)$, an example assembly (not stable) that represents an accepting computation path of M is shown in Fig. 7b. The system created by the reduction also includes the tile set necessary to simulate M in this manner. The set of tiles is disjoint from those used for the binary counter. In the same way, this tile set has the inner tiles (light gray) that perform the computation. These have a strength-1 glue on each side, and border tiles (dark gray). The north border uses a constant number of distinct tile types to ensure that the accept state of the Turing machine must be present in the row below it in order for a stable assembly to be formed.

Production of Computation Assembly. The key question of this system is whether the 70 hands can be utilized to bring together ≤ 70 of the described tiles to produce a computation assembly. In the case where the original instance K-A$_{NTM}$ is true, the answer is 'yes'. For the example provided, 28 hands can be used to arrange the tiles of the binary counter section to form an assembly representing the counting from 0 to 4 (Fig. 7b). Above this, the remaining 42 hands can arrange up to 42 tiles in the Turing machine simulation section in an arrangement that represents an accepting computation path of at most 4 steps (Fig. 7b). The arrangement of binary counter tiles and Turing machine simulation tiles can form a stable assembly if attached to each other (Fig. 7c). Therefore the computation assembly is a producible assembly in a 70-handed system. Note that a computation assembly of size less than 70 can also be produced if there exists an existing computation path strictly less than length 4.

Target Assembly. The target assembly for the H-HAM-UAV instance is an assembly that acts as a "frame" that holds all the tiles previously described (Fig. 8c). Each tile has a designated spot within this assembly that is specified by the frame having the corresponding glues that uniquely identify the tile adjacent to its designated position. Some consideration must be taken in the arrangement of the tiles within this frame to ensure that some extraneous assembly is not built within the frame. Since the tiles that compose the frame have strength-4 glues between them, it is clear that this frame is always built. For every tile in the binary counter section and Turing machine simulation section, there is one spot in the frame that exactly complements its glues, so it is true that each of these tiles can attach to the frame. Thus, the target assembly will always be produced.

5.3   Complexity

Given the membership and reduction overview in Sects. 5.1 and 5.2, respectively, we show the following.

Theorem 4. H-HAM-UAV *is complete for coNEXP.*

Proof. Lemma 5 shows that H-HAM-UAV is in the class coNEXP. The reduction shown is a polynomial time reduction from K-A_{NTM} to H-HAM-UAV taking an instance $P = (M, k)$ to an instance $P' = (\Gamma = (T, h, 4), A)$ where $h = (2^{\lceil \log_2(k) \rceil} + 3) \cdot (k + \lceil \log_2(k) \rceil + 4)$, and $\neg P \iff P'$. It was shown how a true instance of P implies P' is false, through the production of a computation assembly that will never grow into the target assembly.

We now show that the instance P being false implies that the created instance of P' is true. It is clear the target assembly will be produced, but it must be shown that no assembly that is not a subassembly of the target assembly is produced. We first note that the border tiles can not come together alone to form a hollow square. This is because at the points where the border tiles from the binary counter section would meet those from the Turing machine simulation section, there are strength-1 glues (red arrows in Figs. 7a and 7b), meaning the hollow square is not a stable assembly.

From the provided tile types shown, in order to be stable the computation assembly must be enclosed by border tiles (dark gray). Every other tile has only strength-1 glues on every edge, and therefore if the tile were on the border, the assembly would not be stable. Due to a unique glue (shown in green in Fig. 7a), the right border of the binary counter can only be built if there is a 1 to the left of it in the row representing the most significant bit. Therefore, in order to be stable, the binary counter assembly must have counted up to $2^{\lceil \log_2(k) \rceil}$. Thus, $(2^{\lceil \log_2(k) \rceil} + 3) \cdot (\lceil \log_2(k) \rceil + 2)$ of the allotted hands must be used to "hold" the binary counter in place (28 in our example).

This leaves the system with $(2^{\lceil \log_2(k) \rceil} + 3) \cdot (k + 2)$ hands left, which can be used to arrange the Turing machine simulation tiles in a way that can attach to the binary counter. Since only the border tiles of the binary counter assembly attach to the border tiles of the simulation assembly, the simulation assembly must be of the same width ($2^{\lceil \log_2(k) \rceil} + 3$). Thus, it can be at most height $k + 2$. Since one row has to be used for the top border, the simulation can only utilize $k - 1$ rows, i.e., k steps. Due to another unique glue on the top border of the simulation assembly (cyan in Fig. 7b), the tile B_A can only be stable on an assembly if the row below it contains a tile that represents the accept state. Every tile in the Turing machine simulation section must have a matching glue with all of its neighbors. Since every two adjacent rows in the Turing machine simulation share matching glues, the glue encoding enforces that it is a valid transition from one configuration of the Turing machine to another. Therefore, starting from the initial configuration of M, if there does not exist a computation path that accepts in $\leq k$ steps, then there is no way to arrange the $k + 1$ rows in a way that is both stable and has the accept state present. Thus, if the instance P is false, then the only terminal assembly of the created system is the target assembly, so P' is true.   □

Corollary 2. U-H-HAM-UAV *is complete for coNP.*

Proof. Corollary 1 shows that U-H-HAM-UAV is in the class coNP. The problem U-K-A$_{NTM}$ where the parameter k denoting the maximum allotted runtime is encoded in unary is coNP-hard. An equivalent reduction which outputs the same instance $P' = (\Gamma = (T, h, 4), A)$ with the difference that h is encoded in unary is a polynomial time reduction from U-K-A$_{NTM}$ to U-H-HAM-UAV.   □

6 Conclusion

In this paper, we analyzed for the first time two of the most fundamental self-assembly questions in relation to the h-handed model: producibility and unique assembly verification. We proved that producibility is polynomial and UAV is coNP-complete when the number of hands is encoded in unary and coNEXP-complete if it is encoded in binary. Further, we gave a class of shapes that show the power of additional hands by having a provable separation in necessary tile types between shapes. We also showed that with a constant number of tile types, different sized squares are producible depending on the number of hands.

References

1. Adleman, L.M., et al.: Combinatorial optimization problems in self-assembly. In: Proceedings of the 34th Annual ACM Symposium on Theory of Computing, pp. 23–32 (2002)
2. Aggarwal, G., Cheng, Q., Goldwasser, M.H., Kao, M.Y., de Espanes, P.M., Schweller, R.T.: Complexities for generalized models of self-assembly. SIAM J. Comput. **34**(6), 1493–1515 (2005). https://doi.org/10.1137/S0097539704445202
3. Alon, N., Yuster, R., Zwick, U.: Finding and counting given length cycles. Algorithmica **17**, 209–223 (1997). https://doi.org/10.1007/BF02523189

4. Becker, F., Rapaport, I., Rémila, É.: Self-assemblying classes of shapes with a minimum number of tiles, and in optimal time. In: Arun-Kumar, S., Garg, N. (eds.) FSTTCS 2006. LNCS, vol. 4337, pp. 45–56. Springer, Heidelberg (2006). https://doi.org/10.1007/11944836_7
5. Bryans, N., Chiniforooshan, E., Doty, D., Kari, L., Seki, S.: The power of nondeterminism in self-assembly. In: Proceedings of the Twenty-Second Annual ACM-SIAM Symposium on Discrete Algorithms, pp. 590–602. SIAM (2011)
6. Cannon, S., et al.: Two hands are better than one (up to constant factors): self-assembly in the 2HAM vs. aTAM. In: 30th International Symposium on Theoretical Aspects of Computer Science (STACS 2013). Leibniz International Proceedings in Informatics (LIPIcs), vol. 20, pp. 172–184. Schloss Dagstuhl-Leibniz-Zentrum fuer Informatik (2013)
7. Chalk, C., Luchsinger, A., Schweller, R., Wylie, T.: Self-assembly of any shape with constant tile types using high temperature. In: Proceedings of the 26th Annual European Symposium on Algorithms, ESA 2018 (2018)
8. Chalk, C.T., Fernandez, D.A., Huerta, A., Maldonado, M.A., Schweller, R.T., Sweet, L.: Strict self-assembly of fractals using multiple hands. Algorithmica 76(1), 195–224 (2016). https://doi.org/10.1007/s00453-015-0022-x
9. Demaine, E.D., et al.: Staged self-assembly: nanomanufacture of arbitrary shapes with O(1) glues. Nat. Comput. 7(3), 347–370 (2008). https://doi.org/10.1007/s11047-008-9073-0
10. Doty, D.: Randomized self-assembly for exact shapes. SIAM J. Comput. 39(8), 3521–3552 (2010)
11. Doty, D.: Theory of algorithmic self-assembly. Commun. ACM 55(12), 78–88 (2012)
12. Doty, D.: Producibility in hierarchical self-assembly. In: Ibarra, O.H., Kari, L., Kopecki, S. (eds.) UCNC 2014. LNCS, vol. 8553, pp. 142–154. Springer, Cham (2014). https://doi.org/10.1007/978-3-319-08123-6_12
13. Eppstein, D.: Subgraph isomorphism in planar graphs and related problems. J. Graph Algorithms Appl. 3(3), 1–27 (1999). https://doi.org/10.7155/jgaa.00014
14. Fekete, S.P., Schweller, R.T., Winslow, A.: Size-dependent tile self-assembly: constant-height rectangles and stability. In: Elbassioni, K., Makino, K. (eds.) ISAAC 2015. LNCS, vol. 9472, pp. 296–306. Springer, Heidelberg (2015). https://doi.org/10.1007/978-3-662-48971-0_26
15. Kao, M.Y., Schweller, R.: Reducing tile complexity for self-assembly through temperature programming. arXiv preprint cs/0602010 (2006)
16. Kao, M.-Y., Schweller, R.: Randomized self-assembly for approximate shapes. In: Aceto, L., Damgård, I., Goldberg, L.A., Halldórsson, M.M., Ingólfsdóttir, A., Walukiewicz, I. (eds.) ICALP 2008. LNCS, vol. 5125, pp. 370–384. Springer, Heidelberg (2008). https://doi.org/10.1007/978-3-540-70575-8_31
17. Kowalik, Ł: Short cycles in planar graphs. In: Bodlaender, H.L. (ed.) WG 2003. LNCS, vol. 2880, pp. 284–296. Springer, Heidelberg (2003). https://doi.org/10.1007/978-3-540-39890-5_25
18. Kundeti, V., Rajasekaran, S.: Self assembly of rectangular shapes on concentration programming and probabilistic tile assembly models. Nat. Comput. 11(2), 199–207 (2012). https://doi.org/10.1007/s11047-012-9313-1
19. Minev, D., Wintersinger, C.M., Ershova, A., Shih, W.M.: Robust nucleation control via crisscross polymerization of highly coordinated DNA slats. Nat. Commun. 12(1), 1–9 (2021)

20. Patitz, M.J.: An introduction to tile-based self-assembly and a survey of recent results. Nat. Comput. **13**(2), 195–224 (2013). https://doi.org/10.1007/s11047-013-9379-4
21. Rothemund, P.W., Winfree, E.: The program-size complexity of self-assembled squares. In: Proceedings of the Thirty-Second Annual ACM Symposium on Theory of Computing, pp. 459–468 (2000)
22. Schweller, R., Winslow, A., Wylie, T.: Complexities for high-temperature two-handed tile self-assembly. In: Brijder, R., Qian, L. (eds.) DNA 2017. LNCS, vol. 10467, pp. 98–109. Springer, Cham (2017). https://doi.org/10.1007/978-3-319-66799-7_7
23. Schweller, R., Winslow, A., Wylie, T.: Nearly constant tile complexity for any shape in two-handed tile assembly. Algorithmica **81**(8), 3114–3135 (2019). https://doi.org/10.1007/s00453-019-00573-w
24. Schweller, R., Winslow, A., Wylie, T.: Verification in staged tile self-assembly. Nat. Comput. **18**(1), 107–117 (2018). https://doi.org/10.1007/s11047-018-9701-2
25. Summers, S.M.: Reducing tile complexity for the self-assembly of scaled shapes through temperature programming. Algorithmica **63**(1), 117–136 (2012). https://doi.org/10.1007/s00453-011-9522-5
26. Winfree, E.: Algorithmic self-assembly of DNA. Ph.D. thesis, California Institute of Technology, June 1998
27. Woods, D.: Intrinsic universality and the computational power of self-assembly. Philos. Trans. Roy. Soc. A Math. Phys. Eng. Sci. **373**(2046), 20140214 (2015)
28. Woods, D., et al.: Diverse and robust molecular algorithms using reprogrammable DNA self-assembly. Nature **567**(7748), 366–372 (2019). https://doi.org/10.1038/s41586-019-1014-9

Computing with Magnetic Thin Films: Using Film Geometry to Improve Dynamics

Matthew Dale[1,3], Simon O'Keefe[1,3], Angelika Sebald[3],
Susan Stepney[1,3(✉)], and Martin A. Trefzer[2]

[1] Department of Computer Science, University of York, York, UK
susan.stepney@york.ac.uk
[2] Department of Electronic Engineering, University of York, York, UK
[3] York Cross-disciplinary Centre for Systems Analysis, University of York, York, UK

Abstract. Inspired by the nonlinear dynamics of neural networks, new unconventional computing hardware has emerged under the name of physical reservoir computing. In this paradigm, an input-driven dynamical system (the reservoir) is exploited and trained to perform computational tasks. Recent spintronic thin-film reservoirs show state-of-the-art performances despite simplicity in their design. Here, we explore film geometry and show that simple changes to film shape and input location can lead to greater memory and improved performance across various time-series tasks.

1 Introduction

Emerging *in materio* computing systems have the potential for extreme parallelism, ultra-low power consumption, and robustness, making them ideal solutions to challenges in artificial intelligence and robotics [32]. *In materio* computing, in contrast to conventional computing, does not impose a computational model upon the substrate. It performs embodied computation by directly exploiting the natural dynamics of its material composition. This typically involves some excitable media with observable nonlinear behaviour resulting from various intrinsic physical processes. Example substrates include atomic switch networks, skyrmion fabrics, dopant networks, and nanotube composites [6,11,22,25,26,29].

These novel material systems are typically trained or reconfigured through heuristic search rather than explicitly "programmed" – many using the reservoir computing framework [33] – to perform tasks such as classification, time-series prediction, and pattern recognition.

In materio computing harnesses embodied computation, which implies material structure and morphology play a pivotal role. Exploiting morphology provides additional degrees of freedom in design and could allow new material behaviours to emerge, with the potential to fine tune and control the materials' intrinsic dynamics.

I. Kostitsyna and P. Orponen (Eds.): UCNC 2021, LNCS 12984, pp. 19–34, 2021.
https://doi.org/10.1007/978-3-030-87993-8_2

The morphology of *in materio* computers is an area that is largely unexplored. Penty and Tufte [28] have recently provided a glimpse into the potential of manipulating material morphology for computation. They show that the geometries and arrangement of large arrays of nanomagnets (artificial spin ice) can be tuned to discover unique computational states and scalable mechanisms. However, more work is needed to understand how morphology can be fully exploited to engineer desirable computational properties, and potentially optimised for specific applications.

Here, we explore the basic geometry of thin film ferromagnets and how it affects dynamical properties related to reservoir computing. Recent work with these substrates show competitive performances to digital recurrent neural networks across different temporal tasks [8]. The materials used are continuous films where edge effects appear to play an essential role in the computing properties. For example, initial results show film (material reservoir) size affects performance in a way counter to that seen in other reservoir systems, with different aspects of dynamical behaviour changing as size changes.

Based on the assumption that edge effects appear to affect film dynamics, the morphology of the material could provide another level of optimisation and control. Our hypothesis is that breaking the symmetry of the film can lead to more desirable dynamical properties for certain tasks. Here, we test this hypothesis by manipulating the geometry of the film, its size, and the location of inputs, and evaluate what effect it has on the film's intrinsic memory and task performance.

2 Reservoir Computing

Reservoir Computing (RC) [35] is a popular computational framework used to harness and train a variety of input-driven dynamical systems. Reservoir computers comprise several layers: the input, reservoir, readout and output layers. At the input layer, a time-varying signal $\mathbf{u}$ is transformed through the input mapping $\mathbf{W}_{in}$ (typically random) to the next layer: the *reservoir*. The reservoir layer consists of a black-box dynamical system that is perturbed by the inputs. This layer projects the driving input into a high-dimensional spatial-temporal state space $\mathbf{x}$. The readout layer extracts the states $\mathbf{x}$ of the system and performs a linear mapping $\mathbf{W}_{out}$ to the final output nodes $\mathbf{y}$. Training occurs only at the readout layer, typically using supervised learning, to find a mapping $\mathbf{W}_{out}$ that reduces the error between target data $\mathbf{y}^{\mathbf{T}}$ and observed output data $\mathbf{y}$ [23].

The general state update equation for a discrete-time continuous-value reservoir system is as follows:

$$\tilde{\mathbf{x}}(n) = \sigma(b\mathbf{W}_{in}[u_{bias}; \mathbf{u}(n)] + \alpha\mathbf{W}\mathbf{x}(n-1)) \tag{1}$$

$$\mathbf{x}(n) = (1-a)\mathbf{x}(n-1) + a\tilde{\mathbf{x}}(n) \tag{2}$$

$$\mathbf{y}(n) = \mathbf{W}_{out}[u_{bias}; \mathbf{u}(n); \mathbf{x}(n)] \tag{3}$$

where $\mathbf{x}(n)$ is the system state at data point n, u_{bias} is an input bias (usually 1), σ represents the reservoir (e.g., material) function, b is the input scaling, and

α is the internal damping. $\mathbf{W}$ defines an intrinsic function of the system, for example, network weights in an echo state network [23] that transform previous system states. In a physical system, $\mathbf{Wx}$ is defined by properties of the material such as its internal structure. The parameter a is the leak parameter; if $a < 1$, this represents an intermediate leak filter that allows previous state values to "leak" into current state values. Varying the parameter a can help match the internal timescales of the system to the characteristic timescale of the task.

There are many variations of how to construct and train reservoirs, including simplified topologies [10] and delay-line systems [3]. Different optimisation techniques are also used, such as Bayesian optimisation and artificial evolution [2,5,9].

The generalisability of the RC framework and its transferable techniques is currently driving an analogue computing revival. A sub-field of RC, referred to as physical RC, harnesses the nonlinear properties of physical dynamical systems to perform machine learning. A variety of reservoir systems have been proposed including optical, electronic, spintronic, quantum, and mechanical (see review [33]).

Spintronic reservoirs in particular offer many desirable features for constructing new computing systems, including intrinsic memory, nanometer-size, ultra low-power consumption, high response frequencies (GHz–THz), and easy integration with current CMOS technologies [21,31,34]. Spintronic reservoirs exploit the intrinsic spin of electrons and their associated magnetic moment. A variety of magnetic behaviours have been exploited, including magnetic domain walls, nonlinear oscillation and spin waves [15,16,29]. These reservoir systems are still in early development, and work on scaling them efficiently is underway.

3   Magnetic Reservoir

The simulated physical reservoir system used in this work exploits the input-driven spin dynamics of ferromagnetic thin films, following from previous work [8]. The inherent volatility and nonlinear dynamics of precessing spins provide a temporal-spatial embedding of different magnetic states to perform computation [24]. Figure 1 outlines the basic reservoir representation and system interface. The reservoir representation defines discrete 'macro-cells' containing multiple spins, for the purpose of micromagnetic simulation and input-output locations.

The thin film shape creates a highly structured crystal lattice. Driving signals (localised external magnetic fields) propagate through the lattice via the intrinsic coupling between spins. The coupling of nearest neighbours is determined by physical and material properties of exchange, anisotropy, and dipole fields. Exchange interactions dominate over short length-scales, meaning that macro-cells have finite temporal and spatial correlations over the total sample size.

The simulated film is divided into a grid of $(5\,\text{nm}) \times (5\,\text{nm})$ macro-cells, where each cell can be driven by a local time-varying magnetic field. The strength of

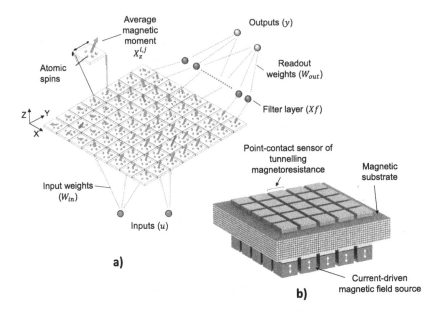

Fig. 1. Magnetic platform: (a) Layout of reservoir system including input mapping (W_{in}), leak filter (Xf) and readout (W_{out}) layers. In simulation, atomistic detail is coarse grained into macro-cells to model the film's micro-magnetic behaviour. (b) Possible implementation of a physical device, with film and interfacing. The film is sandwiched between input and readout contacts, and is perturbed by current-driven, localised, external magnetic fields. The state of the film is recorded using tunnelling magnetoresistance.

the magnetic field is determined by the weighted input mapping $\mathbf{W}_{in}$, which is connected to the data source and a bias. The magnetic field is induced by an electrical current and applied in the film's z-direction.

The film's magnetic state is recorded via nano-contacts measuring the local tunnelling magneto-resistance (TMR). The average magnetisation within each macro-cell is represented by a state vector, used to perform training in the readout layer.

The atomistic simulator VAMPIRE [13] is used to simulate the system dynamics. The system is simulated micro-magnetically where atomistic detail is coarse grained into macro-cells. The energetics of the micromagnetic system are described using a spin Hamiltonian neglecting non-magnetic contributions and given by:

$$\mathcal{H}_{eff} = \mathcal{H}_{app} + \mathcal{H}_{ani} + \mathcal{H}_{exc} + \mathcal{H}_{dip} \tag{4}$$

where $\mathcal{H}_{app}$ is the applied field, $\mathcal{H}_{ani}$ is the anisotropy field, $\mathcal{H}_{exc}$ is the intergranular exchange, and $\mathcal{H}_{dip}$ is the dipole field. To model time-dependent behaviour, the atomistic Landau-Lifshitz-Gilbert (LLG) equation is applied. Full details about the physical model and simulation are given in [8].

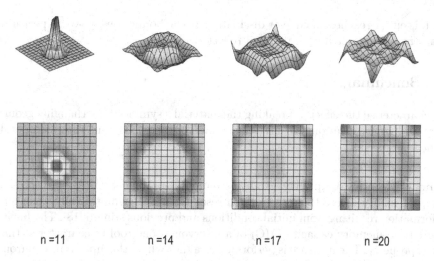

Fig. 2. Micromagnetic response to an input pulse in the z-direction with side and top-down views. The input is applied to the centre cell of a 14×14 grid. The magnetic state of the film is shown at short intervals after the pulse is applied at $n = 10$.

Here, we focus on thin films of cobalt (Co). Previous work [8] shows Co films can exhibit a broad dynamical range and consistently perform well across different benchmark tasks compared to other ferromagnetic materials. Simulation parameters describing the material properties such as exchange and anisotropy for Co are given in [8].

To generate useful computational dynamics, the input frequency has to closely match the internal timescales of the system, for example, the relaxation and precession of spins. The chosen input frequency of our system is 10 ps/100 GHz: each data sample n is applied and held for 10 ps. Previous experiments show a 10 ps timescale provides suitable settling and response times to exhibit useful dynamics for computation. However, it may be possible to operate at slower speeds (1–5 GHz), and even quasi-statically.

In simulation, film thickness is set to approximate 2D films. Films are simulated at zero Kelvin in order to observe only pure magnetic behaviour. Higher temperatures lead to thermal noise which typically degrades performance. In [8], we show that some films can nevertheless perform competitively to digital reservoirs at temperatures up to 100K.

Here, the input mapping, readout layer and training are computed externally. This includes a filter layer $\mathbf{X}_f(n)$ that applies a simple one-step low pass filter controlled by the leak rate parameter a (see [8] for details), providing an additional parameter to match the timescales of the system and task.

Figure 2 demonstrates the film's micromagnetic response to an input pulse applied at its centre, after successive time intervals. When perturbed at time $n = 10$, a propagating wave is initiated and travels outwards across the film until it either reaches an edge or dissipates. At the boundaries, waves experience reflective effects, leading to damped reflected waves.

4   Benchmarks

To characterise the effect of breaking the material's symmetry by changing geometry and input location, we investigate memory capacity and three temporal benchmark tasks.

Linear Memory Capacity. To solve temporal problems, a reservoir needs a fading memory where the input driven reservoir must asymptotically wash out information resulting from initial conditions and previous stimuli [18]. The *linear* short-term memory capacity (MC) of a reservoir is one tool to characterise this basic property. The measure is performed as a task where the input is drawn from a random uniform distribution and injected into the reservoir. The readout is then trained to recover previous inputs $u(n-k)$ separated into k outputs where $k = 1, 2, 3 \ldots, 2N$; N is typically the number of nodes or observable states. Memory capacity is measured as how much variance of the delayed input can be recovered by the trained reservoir outputs, summed over all delays:

$$MC = \sum_{k=1}^{2N} MC_k = \sum_{k=1}^{2N} \frac{cov^2(u(n-k), y(n))}{\sigma^2(u(n))\sigma^2(y(n))} \tag{5}$$

A total of 1,000 values are generated and split into: 500 training and 500 test. The first 50 values of each sub-set are discarded as an initial washout period.

Nonlinear Channel Communication. The task replicates the equalisation of a wireless communication channel, as described in [12,19,27,30]. The objective is to recover the original symbol sequence $d(n)$ which is modulated and transmitted as $q(n)$ and received as $u(n)$ – a corrupted version of $d(n)$. The reservoir is trained and tasked to recover $d(n-2)$ when $u(n)$ is presented at the input.

The data for this task is as follows: the original symbol sequence $d(n)$ is generated from a uniform random distribution of values $d(n) \in \{-3, -1, +1, +3\}$. $d(n)$ is transformed to create the transmitting signal $q(n)$ through a filter:

$$\begin{aligned} q(n) = {} & 0.08d(n+2) - 0.12d(n+1) + d(n) + 0.18d(n-1) \\ & - 0.1d(n-2) + 0.091d(n-3) - 0.05d(n-4) \\ & + 0.04d(n-5) + 0.03d(n-6) + 0.01d(n-7) \end{aligned} \tag{6}$$

To corrupt the signal, a nonlinear transformation is applied to $q(n)$ to produce the signal $u(n)$:

$$u(n) = q(n) + 0.036q(n)^2 - 0.011q(n)^3 + v(n) \tag{7}$$

where $v(n)$ represents i.i.d. Gaussian noise with zero mean and adjusted in power to yield signal-to-noise ratio of 28 dB. Following [30], the input $u(n)$ signal is shifted by $+30$.

The error is calculated using the Symbol Error Rate (SER) representing the fraction of incorrect symbols. To calculate SER, an estimator $\hat{d}(n)$ replaces the reservoir output $y(n)$ with the closest discretised value $\{-3, -1, +1, +3\}$.

A total of 8,000 values are generated and split into: 2,000 training, 3,000 validation, and 3,000 test. The first 200 values of each sub-set are discarded as an initial washout period.

NARMA-10. The NARMA (nonlinear autoregressive moving average) task [4] evaluates a reservoir's ability to model a 10-th order non-linear dynamical system. The task contains both non-linearity and a long-term dependency created by the 10-th order time-lag. The task is to predict the output $y(n + 1)$ given by Eq. (8) when supplied with $u(n)$ from a uniform distribution of interval $[0, 0.5]$. For the 10-th order systems $\alpha = 0.3$, $\beta = 0.05$, $\delta = 10$ and $\gamma = 0.1$.

$$y(n+1) = \alpha y(n) + \beta y(n) \left(\sum_{i=0}^{\delta} y(n-i) \right) + 1.5u(n-\delta)u(n) + \gamma \tag{8}$$

A total of 5,000 values are generated and split into: 3,000 training, 1,000 validation, and 1,000 test. The first 50 values of each sub-set are discarded as an initial washout period.

IPIX Radar. The IPIX radar dataset is popular task in RC and has been applied to various reservoir types in the literature [12,27,30,36].

IPIX radar is a noisy prediction task based on real-world data collected by the McMaster University IPIX radar[1]. The target signal is sea clutter data recorded as radar backscatter from an ocean surface under low sea state conditions. The signal has two dimensions: the in-phase and in-quadrature outputs (I and Q) of the radar demodulator. The task requires the successful prediction of both dimensions, i.e. the task is a two-input, two-output problem.

The task is to predict 5 steps ahead $y(n) = u(n + 5)$ when $u(n)$ is presented at the reservoir input.

Following [30], a total of 2,000 values are generated and split into: 800 training, 500 validation, and 700 test. The first 100 values of each sub-set are discarded as an initial washout period.

---

[1] Accessible from: http://soma.ece.mcmaster.ca/ipix/ (2021).

5 Film Geometry

5.1 Experimental Setup

To explore basic film geometry, we investigate rectangular films with various heights and widths. The objective is to break the input symmetry by varying the distances between the input source and the edges by adjusting the shape. Rectangular films have less symmetry than square films, and can be configured in different ratios from square-like to a one dimensional (one macro-cell wide) line or wire.

In addition, we also investigate what effect the height:width ratio has as the number of cells increases from 36 (6^2) cells up to 196 (14^2). Previous work [8] shows that gains in performance tail off as reservoir size increases. It is likely a threshold in size exists where bulk material properties begin to dominate, as edge effects are diminished or become negligible. Therefore, the richness produced by reflections and interference dies off with the probability of interactions decreasing as distance increases.

To compare fairly, films with different ratios must have equal areas, to maintain an equal number of cells. Access to more states can be beneficial to the readout during training and skew comparisons. Height and width are adjusted accordingly to maintain a specific number of cells, which in turn limits the number of ratios that can be tested.

To simplify experiments and exclude effects from multiple inputs, a single input is applied to the film's centre macro-cell. This input carries the task input with a single adjustable input gain b (see Eq. 1). The readout and training has access to all available cell states.

In the following experiments, cobalt reservoirs are configured using three parameters with decimal values each taken from a random uniform distribution: input scaling ($-1 < b \leq 1$), intrinsic material damping ($0.001 < \alpha \leq 1$), and leak rate ($0 < a \leq 1$). From previous work [8], we know a low damping tends to increase memory; input scaling can be tuned to improve memory further using low values, or increase non-linearity with high values. Leak rate on the other hand tends to vary depending on the task dynamics.

5.2 Effect of Height:Width Ratio on Memory Capacity

This experiment investigates what effect the height:width ratio has on memory. Applying the linear memory capacity measure (MC), multiple height:width ratios are tested for each film size. To assess the overall effect, independent of reservoir parameters and size, the same 250 random reservoir parameters (b, α and a) are applied to every ratio and size.

Figure 3 shows that memory varies with respect to both size and ratio. In terms of film size, the average and maximum memory capacities of small-to-medium reservoirs is large compared to the bigger reservoir films. As the height:width ratio decreases, (as films become elongated), the mean memory capacity

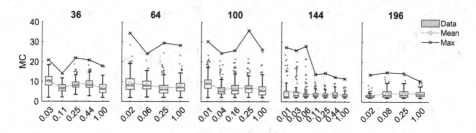

Fig. 3. Memory capacity (MC) at various film sizes (number of cells) and different ratios. Ratio (height:width) is given along the x-axis. Each box plot shows the MC of 250 reservoirs, covering a range of parameter settings (input scaling, damping, leak rate). The same 250 reservoir parameter sets are used for each size and ratio. The mean and max MC are highlighted.

tends to increase relative to the default square (ratio of 1). This pattern is consistent across all sizes, except for the largest reservoir with 196 cells.

This experiment highlights three key points: i) using a basic setup with random parameters and a single input, the mean memory capacity tends to decrease as size increases, ii) an elongated film increases memory capacity, and iii) above a certain film size, memory capacity tends to plateau. Memory, however, is only one factor that affects reservoir performance, and how much memory is required depends on the task.

5.3   Effect of Height:Width Ratio on Task Performance

This experiment measures the effect film height:width ratio has on task performance. To assess the overall effect, the same 250 random reservoir parameters as above are reused and evaluated for each ratio. The task input is again transformed into a single local magnetic field applied at the centre of the film.

Figure 4 shows performance on the NARMA-10, IPIX radar, and Nonlinear channel communication (NCC) tasks. For each task, performance improves when rectangular films are used. The films that tend to work best have a ratio of 0.25 or 0.44. These typically show a statistically significant decrease in the average normalised mean square error (NMSE) or symbol error rate (SER), and produce the lowest errors at each size. In general, a ratio that is too long and thin tends to be detrimental to performance and increases error. This suggests increasing memory capacity alone does not improve performance. However, other properties such as nonlinearity of the films may have also changed which may now strongly affect task performance.

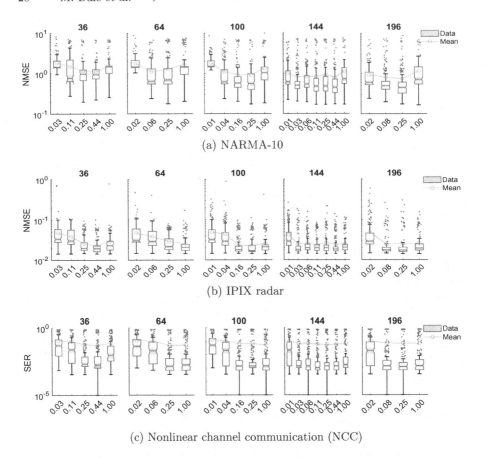

(a) NARMA-10

(b) IPIX radar

(c) Nonlinear channel communication (NCC)

Fig. 4. Task error at each size and ratio. Each box plot shows errors for 250 random reservoirs. If box plot notches do not overlap, there is a 95% confidence that the medians differ; we see that ratio has a significant effect on error. The mean is highlighted, showing the error distribution is skewed towards smaller errors, i.e. smaller errors are more common when parameters are chosen from a uniform random distribution. For the NCC task, an $SER = 10^{-5}$ is used to represent a zero error in order to plot error logarithmically.

6   Breaking Input Symmetry

6.1   Experimental Setup

In this section, we investigate the effect of input location. An input that is off-centre can be used to break the symmetry of a square film without having to change film dimensions. To test this, we define four quadrants and apply a single input towards the centre of each quadrant; all quadrants are tested on the assumption that the film may not be symmetric in terms of magnetic behaviour in all directions.

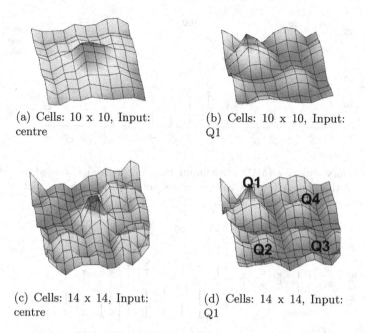

(a) Cells: 10 x 10, Input: centre

(b) Cells: 10 x 10, Input: Q1

(c) Cells: 14 x 14, Input: centre

(d) Cells: 14 x 14, Input: Q1

Fig. 5. Visualisation of input symmetry using wave interference at two film sizes. A sine wave is applied in the z-direction at the film centre (a and c) and in quadrant Q1 (b and d). The plot shows the summation of the absolute values in each cell (in the z-direction) over 1000 time steps. Peaks and troughs outline underlying symmetry.

The objective here is to break the input symmetry by varying the distances between the input source and the edges by moving the input, rather than by adjusting the shape. In Fig. 5, we visualise the input symmetry of square films by observing wave interference. To do this, we apply a sine wave and plot the summation of the absolute values over time. This shows the corresponding wave peaks and troughs produced by reflections at the edges. From this observation, we see that the film's behaviour to stimulus is roughly symmetric given the visual symmetry in peaks and troughs; however, small deviations are present.

6.2 Effect of Input Location on Memory Capacity

This experiment to test input symmetry measures the effect input location has on memory capacity. Figure 6 shows that an off-centre input significantly changes the average memory capacity and tends to increase the maximum memory capacity. As found previously, memory capacity reduces as film size increases. The effect of changing input location is therefore more prominent at smaller film sizes. The results suggest that moving the input looks qualitatively similar to 'stretching' the film.

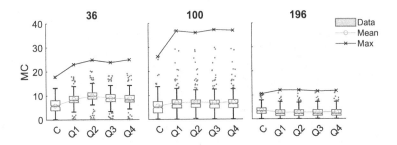

Fig. 6. Memory capacity (MC) when input location (x-axis) is moved into different quadrants of the film. A total of 250 reservoirs are given in each box plot. The mean and maximum MC are also provided.

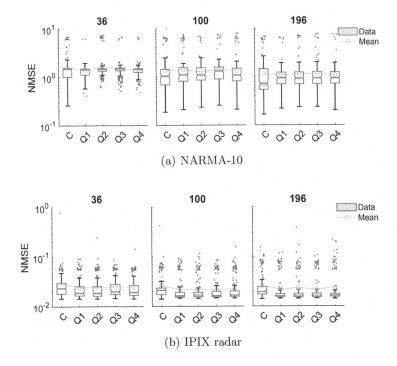

Fig. 7. Normalised mean squared error (NMSE) on NARMA-10 and IPIX Radar tasks when the single driving input is applied to different quadrants (Q1–4). Each box plot shows errors for 250 random reservoirs.

6.3 Effect of Input Location on Task Performance

This experiment evaluates task performance, and shows that the similarity to 'stratching' does not hold in this case. Figure 7 shows the average error is similar, or worse, for the NARMA-10 task at different input locations, but improves for the IPIX task.

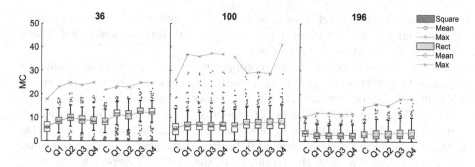

Fig. 8. Memory capacity (MC) when film geometry and input location are changed. Square films (left, blue) are compared to rectangular films (right, purple) with a ratio of 0.25. Each input location tested is given along the x-axis, i.e. C = centre and quadrants Q1–4. (Color figure online)

In summary, moving the input location alters the basic computational properties of the film, i.e. improving memory, but whether this translates into improved performance depends on the task.

6.4 Effect of Geometry and Input Location on Memory Capacity

This experiment investigates the effect of varying both geometry and input location on memory capacity. Altering each individually shows memory capacity can increase. Now we test whether combining both can increase memory capacity further.

The film height:width ratio is set to 0.25, and a single input is applied to a quadrant. The same 250 random configurations from previous experiments are used to assess memory capacity. Figure 8 compares the original memory capacity results of the square film (left, blue) to the rectangular film (right, purple) at different film sizes. When applying an off-centre input to the rectangular film, the average memory capacity increases further. Therefore the same random configurations tend to produce higher memory capacities. At the largest film size, we also find more configurations with larger memory capacities than before.

In summary, moving the input location continues to increase memory capacity when the geometry of the film is also changed. This suggests boundary effects and distance to the edges strongly influence the dynamical response, particularly for smaller reservoir films.

7 Conclusion

Harnessing physical processes such as wave propagation for computation in excitable media could lead to many efficient novel computing platforms [1,7, 14,17,20]. Exploiting and controlling aspects of wave-like propagation such as speed, modulation and reflections, which are susceptible to the system's morphology, can produce interference patterns, phase distortions and signal delays exploitable for information processing.

Here, we show that the dynamical properties of magnetic reservoir computing films can be partially controlled using the film's geometry. For example, creating long and narrow rectangular films increases memory capacity for a given film area. Whether these changes correspond to improvements in task performance largely depend on the task. However, in general, we find that rectangular films with height:width ratios around 0.25 often show significant improvements. We also find that breaking the symmetry of the film with respect to the input position can lead to greater memory capacity, but again this does not lead to universal improvements in task performances. These results suggest a strong correlation between the dynamical response and boundary effects, and highlight the requirement to match reservoir dynamics to the dynamics of task.

There are numerous other ways to manipulate and break the symmetry of magnetic reservoir films, including multiple driving inputs, bias (static) inputs, and various shapes with different lines of symmetry. In [8], artificial evolution is used to decide the location, strength, and number of inputs and biases. This results in further improvements in performance, for example, on the NARMA-10 task. Allowing evolution to manipulate the film's geometry could lead to films better tailored to specific tasks, and will be explored in future work.

The inherent spatial and temporal properties of the investigated system may limit computation to a local area; this property could be a desirable feature for large scale systems [17]. Reservoir scaling, both physically and in terms of computing capabilities, could be achieved by combining many smaller material systems, such as those investigated here, with different but complementary dynamics, rather than large monolithic reservoir systems. A drive towards multi-reservoir computing architectures could significantly benefit this particular reservoir system.

Acknowledgments. Thanks to Richard Evans and Sarah Jenkins for help with the VAMPIRE simulator. All experiments were carried out using the Viking Cluster, a high performance compute facility provided by the University of York. This work was funded by the SpInspired project, EPSRC grant EP/R032823/1.

References

1. Adamatzky, A., Costello, B.D.L., Asai, T.: Reaction-Diffusion Computers. Elsevier, Amsterdam (2005)
2. Antonik, P., Marsal, N., Brunner, D., Rontani, D.: Bayesian optimisation of large-scale photonic reservoir computers. Cogn. Comput. 1–9 (2021). https://doi.org/10.1007/s12559-020-09732-6
3. Appeltant, L., et al.: Information processing using a single dynamical node as complex system. Nat. Commun. **2**, 468 (2011). https://doi.org/10.1038/ncomms1476
4. Atiya, A.F., Parlos, A.G.: New results on recurrent network training: unifying the algorithms and accelerating convergence. IEEE Trans. Neural Netw. **11**(3), 697–709 (2000). https://doi.org/10.1109/72.846741

5. Bala, A., Ismail, I., Ibrahim, R., Sait, S.M.: Applications of metaheuristics in reservoir computing techniques: a review. IEEE Access **6**, 58012–58029 (2018). https://doi.org/10.1109/ACCESS.2018.2873770

6. Chen, T., et al.: Classification with a disordered dopant-atom network in silicon. Nature **577**(7790), 341–345 (2020). https://doi.org/10.1038/s41586-019-1901-0

7. Chumak, A.V., Vasyuchka, V.I., Serga, A.A., Hillebrands, B.: Magnon spintronics. Nat. Phys. **11**(6), 453–461 (2015). https://doi.org/10.1038/nphys3347

8. Dale, M., et al.: Reservoir computing with thin-film ferromagnetic devices. arXiv preprint arXiv:2101.12700 (2021)

9. Dale, M., Miller, J.F., Stepney, S., Trefzer, M.A.: Evolving carbon nanotube reservoir computers. In: Amos, M., Condon, A. (eds.) UCNC 2016. LNCS, vol. 9726, pp. 49–61. Springer, Cham (2016). https://doi.org/10.1007/978-3-319-41312-9_5

10. Dale, M., O'Keefe, S., Sebald, A., Stepney, S., Trefzer, M.A.: Reservoir computing quality: connectivity and topology. Nat. Comput. **20**(2), 205–216 (2020). https://doi.org/10.1007/s11047-020-09823-1

11. Dale, M., Stepney, S., Miller, J.F., Trefzer, M.: Reservoir computing in materio: a computational framework for in materio computing. In: IJCNN 2017, pp. 2178–2185. IEEE (2017). https://doi.org/10.1109/IJCNN.2017.7966119

12. Duport, F., Schneider, B., Smerieri, A., Haelterman, M., Massar, S.: All-optical reservoir computing. Opt. Express **20**(20), 22783–22795 (2012). https://doi.org/10.1364/OE.20.022783

13. Evans, R.F.L., Fan, W.J., Chureemart, P., Ostler, T.A., Ellis, M.O.A., Chantrell, R.W.: Atomistic spin model simulations of magnetic nanomaterials. J. Phys. Condens. Matter **26**(10) (2014). https://doi.org/10.1088/0953-8984/26/10/103202

14. Fernando, C., Sojakka, S.: Pattern recognition in a bucket. In: Banzhaf, W., Ziegler, J., Christaller, T., Dittrich, P., Kim, J.T. (eds.) ECAL 2003. LNCS (LNAI), vol. 2801, pp. 588–597. Springer, Heidelberg (2003). https://doi.org/10.1007/978-3-540-39432-7_63

15. Grollier, J., Querlioz, D., Camsari, K., Everschor-Sitte, K., Fukami, S., Stiles, M.D.: Neuromorphic spintronics. Nat. Electron. **3**(7), 360–370 (2020). https://doi.org/10.1038/s41928-019-0360-9

16. Ichimura, T., Nakane, R., Tanaka, G., Hirose, A.: A numerical exploration of signal detector arrangement in a spin-wave reservoir computing device. IEEE Access **9**, 72637–72646 (2021). https://doi.org/10.1109/ACCESS.2021.3079583

17. Izhikevich, E.M., Hoppensteadt, F.C.: Polychronous wavefront computations. Int. J. Bifurcat. Chaos **19**(05), 1733–1739 (2009). https://doi.org/10.1142/S0218127409023809

18. Jaeger, H.: Short term memory in echo state networks. GMD-Forschungszentrum Informationstechnik (2001)

19. Jaeger, H., Haas, H.: Harnessing nonlinearity: predicting chaotic systems and saving energy in wireless communication. Science **304**(5667), 78–80 (2004). https://doi.org/10.1126/science.1091277

20. Jakubowski, M.H., Steiglitz, K., Squier, R.: Computing with solitons: a review and prospectus. In: Adamatzky, A. (ed.) Collision-Based Computing, pp. 277–297. Springer, London (2002). https://doi.org/10.1007/978-1-4471-0129-1_10

21. Kurenkov, A., Fukami, S., Ohno, H.: Neuromorphic computing with antiferromagnetic spintronics. J. Appl. Phys. **128**(1), 010902 (2020). https://doi.org/10.1063/5.0009482

22. Lilak, S., et al.: Spoken digit classification by in-materio reservoir computing with neuromorphic atomic switch networks. arXiv preprint arXiv:2103.12835 (2021)

23. Lukoševičius, M.: A practical guide to applying echo state networks. In: Montavon, G., Orr, G.B., Müller, K.-R. (eds.) Neural Networks: Tricks of the Trade. LNCS, vol. 7700, pp. 659–686. Springer, Heidelberg (2012). https://doi.org/10.1007/978-3-642-35289-8_36

24. Macià, F., Kent, A.D., Hoppensteadt, F.C.: Spin-wave interference patterns created by spin-torque nano-oscillators for memory and computation. Nanotechnology **22**(9), 095301 (2011). https://doi.org/10.1088/0957-4484/22/9/095301

25. Miller, J.F., Harding, S.L., Tufte, G.: Evolution-in-materio: evolving computation in materials. Evol. Intell. **7**(1), 49–67 (2014). https://doi.org/10.1007/s12065-014-0106-6

26. Nakajima, K., Hauser, H., Li, T., Pfeifer, R.: Information processing via physical soft body. Sci. Rep. **5**(1), 1–11 (2015). https://doi.org/10.1038/srep10487

27. Paquot, Y., et al.: Optoelectronic reservoir computing. Sci. Rep. **2** (2012). https://doi.org/10.1038/srep00287

28. Penty, A., Tufte, G.: A representation of artificial spin ice for evolutionary search. In: ALIFE 2021: The 2021 Conference on Artificial Life. MIT Press (2021). https://doi.org/10.1162/isal_a_00436

29. Pinna, D., Bourianoff, G., Everschor-Sitte, K.: Reservoir computing with random skyrmion textures. Phys. Rev. Appl. **14**, 054020 (2020). https://doi.org/10.1103/PhysRevApplied.14.054020

30. Rodan, A., Tiňo, P.: Simple deterministically constructed recurrent neural networks. In: Fyfe, C., Tino, P., Charles, D., Garcia-Osorio, C., Yin, H. (eds.) IDEAL 2010. LNCS, vol. 6283, pp. 267–274. Springer, Heidelberg (2010). https://doi.org/10.1007/978-3-642-15381-5_33

31. Romera, M., et al.: Vowel recognition with four coupled spin-torque nano-oscillators. Nature **563**(7730), 230–234 (2018). https://doi.org/10.1038/s41586-018-0632-y

32. Stepney, S., Rasmussen, S., Amos, M.: Computational Matter. Springer, Cham (2018). https://doi.org/10.1007/978-3-319-65826-1

33. Tanaka, G., et al.: Recent advances in physical reservoir computing: a review. Neural Netw. **115**, 100–123 (2019). https://doi.org/10.1016/j.neunet.2019.03.005

34. Torrejon, J., et al.: Neuromorphic computing with nanoscale spintronic oscillators. Nature **547**(7664), 428–431 (2017). https://doi.org/10.1038/nature23011

35. Verstraeten, D., Schrauwen, B., d'Haene, M., Stroobandt, D.: An experimental unification of reservoir computing methods. Neural Netw. **20**(3), 391–403 (2007). https://doi.org/10.1016/j.neunet.2007.04.003

36. Xue, Y., Yang, L., Haykin, S.: Decoupled echo state networks with lateral inhibition. Neural Netw. **20**(3), 365–376 (2007). https://doi.org/10.1016/j.neunet.2007.04.014

Robust Real-Time Computing with Chemical Reaction Networks

Willem Fletcher[1], Titus H. Klinge[2] (iD), James I. Lathrop[3], Dawn A. Nye[3(✉)] (iD), and Matthew Rayman[3]

[1] Carleton College, Northfield, MN, USA
fletcherw@carleton.edu
[2] Drake University, Des Moines, IA, USA
titus.klinge@drake.edu
[3] Iowa State University, Ames, IA, USA
{jil,omacron,marayman}@iastate.edu

Abstract. Recent research into analog computing has introduced new notions of computing real numbers. Huang, Klinge, Lathrop, Li, and Lutz defined a notion of computing real numbers in real-time with chemical reaction networks (CRNs), introducing the classes $\mathbb{R}_{LCRN}$ (the class of all Lyapunov CRN-computable real numbers) and $\mathbb{R}_{RTCRN}$ (the class of all real-time CRN-computable numbers). In their paper, they show the inclusion of the real algebraic numbers $ALG \subseteq \mathbb{R}_{LCRN} \subseteq \mathbb{R}_{RTCRN}$ and that $ALG \subsetneq \mathbb{R}_{RTCRN}$ but leave open where the inclusion is proper. In this paper, we resolve this open problem and show $ALG = \mathbb{R}_{LCRN} \subsetneq \mathbb{R}_{RTCRN}$. However, their definition of real-time computation is fragile in the sense that it is sensitive to perturbations in initial conditions. To resolve this flaw, we further require a CRN to withstand these perturbations. In doing so, we arrive at a discrete model of memory. This approach has several benefits. First, a bounded CRN may compute values approximately in finite time. Second, a CRN can tolerate small perturbations of its species' concentrations. Third, taking a measurement of a CRN's state only requires precision proportional to the exactness of these approximations. Lastly, if a CRN requires only finite memory, this model and Turing machines are equivalent under real-time simulations.

Keywords: Real time · Chemical reaction networks · Robustness · Analog computing

1 Introduction

Over the last few decades, many theories of molecular computing have emerged. These theories help inform experimental research and help explore the boundaries of nanoscale computation. Some models of molecular programming are structural, such as algorithmic self-assembly [8,9]; some models are amorphous,

This research supported in part by NSF grants 1900716 and 1545028.

I. Kostitsyna and P. Orponen (Eds.): UCNC 2021, LNCS 12984, pp. 35–50, 2021.
https://doi.org/10.1007/978-3-030-87993-8_3

such as chemical reaction networks [4,15]; and some models combine these to characterize more complex interactions [5,14]. Since molecular programming is a relatively new field, many open problems exist concerning the computational limits of these models.

Investigating the complexity of computing real numbers in computational models has historically significant roots. In Turing's famous 1936 paper [16], he defined a real number to be computable if its "expression as a decimal is calculable with finite means." Real numbers can also be classified according to how efficiently they can be computed by a Turing machine. For example, rational numbers are efficiently computable because their recurring decimal pattern can be produced in real time—even by a finite automaton. More formally, a number $\alpha \in \mathbb{R}$ is *real-time computable by a Turing machine* if n bits of its fractional component can be produced in $O(n)$ time. Many transcendental numbers are known to be real-time computable, but surprisingly, no irrational algebraic number is known to be real-time computable. In fact, in 1965, Hartmanis and Stearns conjectured that if $\alpha \in \mathbb{R}$ is real-time computable by a Turing machine, then it is either rational or transcendental [10].

Recent research into analog computing introduced new notions of computing real numbers. Bournez et al. introduced the notion of computing a real number *in the limit* with a general purpose analog computer (GPAC) [1]. To compute $\alpha \in \mathbb{R}$ "in the limit," a designated variable $x(t)$ must satisfy $\lim_{t\to\infty} x(t) = \alpha$. Computing real numbers in this way has also been investigated in population protocols [2] and chemical reaction networks (CRNs) [11]. Huang et al. defined a number $\alpha \in \mathbb{R}$ to be *real-time computable by chemical reaction networks*, written $\alpha \in \mathbb{R}_{\text{RTCRN}}$, if there exists a CRN with integral rate constants and a designated species X such that, if all species concentrations are initialized to zero, then $x(t)$ converges to α exponentially quickly [12]. This means that after n seconds, the concentration of X is within 2^{-n} of α, so the CRN gains one bit of accuracy every second. Huang et al. also required that all species concentrations be bounded to avoid the so-called *Zeno paradox* of performing an infinite amount of computation in finite time using a fast-growing catalyst species [3]. When this restriction is lifted, the measure of time is no longer linear but rather a function of arc length. In this sense, no power is lost via imposing a boundedness requirement. Further, it eliminates the undesirable Zeno paradox from the model.

A key aspect of Huang et al.'s definition of $\mathbb{R}_{\text{RTCRN}}$ is the requirement that the CRN be initialized to *all zeros*, prohibiting any encoding of α in the initial condition of the CRN. The authors showed that $e, \pi \in \mathbb{R}_{\text{RTCRN}}$, leveraging the fact that the initial condition is *exact*. However, these constructions fail if their initial conditions are perturbed by any $\epsilon > 0$. Huang et al. also defined a subfield of $\mathbb{R}_{\text{RTCRN}}$ they called *Lyapunov CRN-computable real numbers*, written $\mathbb{R}_{\text{LCRN}}$. The definition of $\mathbb{R}_{\text{LCRN}}$ is similar to $\mathbb{R}_{\text{RTCRN}}$ except with the additional constraint that the terminating state of the CRN must be an *exponentially stable equilibrium point*. Since an exponentially stable equilibrium point is *attracting*, any initial condition within its basin of attraction will converge exponentially quickly to it. As a result, any $\alpha \in \mathbb{R}_{LCRN}$ can be computed even in the presence of bounded perturbations to initial conditions. Huang et al. also proved that $ALG \subseteq \mathbb{R}_{LCRN} \subseteq \mathbb{R}_{RTCRN}$ where ALG is the set of algebraic real numbers. The authors left as an open problem which of these inclusions is strict.

An additional consequence of computing a real number α "in the limit" with CRNs is that recovering the bits of α is difficult. Even if we produce α *exactly* in the concentration of a species X, we cannot read its individual bits without an infinitely precise measurement device. Alternatively, if a CRN produced the bits of α as a sequence of measurable memory states, then the bits can be read even with imperfect measurements.

Another limitation of this method of computation is in implementation. The concentration of a species in a solution containing a CRN is ultimately determined by the discrete, integral count of the species. This places a countable limit on the number of "exact" values a concentration can achieve even when a CRN is otherwise perfectly initialized and executed. In the mass action model, we often wave away this issue precisely because we do not have an infinitely precise measurement device. This does, however, somewhat obviate the point of being able to calculate values precisely. In fact, previous results concerning CRNs frequently abuse this hand waving to reach theorems that are true of the mass action kinematics but not of the reality it models. Instead, a more reasonable question to ask is what values can we calculate robustly, quickly, and approximately.

In this paper, we show $ALG = \mathbb{R}_{LCRN} \subsetneq \mathbb{R}_{RTCRN}$ to resolve the open problem stated above. This fully characterizes what values we may compute robustly and quickly; however, this definition of computation yet suffers from the inherent flaws described above. To resolve this weakness of the model, we acknowledge these limitations and loosen the definition of computation to accept approximate results. To do so, we only require that a CRN produces approximations of numbers in the sense that an open interval around a concentration α is in an equivalence class with α itself. This approach has three major benefits. First, a bounded CRN may compute not only a single value in finite time but also a sequence of values. Second, a CRN can tolerate small perturbations of its species' concentrations (and potentially other parameters). Third, taking a measurement of a CRN's state only requires precision proportional to the smallest of these intervals.

If we then fix a collection of these intervals into collection of memory maps for a CRN's species and allow it to compute their corresponding memory states in sequence, we obtain a discrete model characterizing a robust chemical computer. Indeed, given, in this sense, a robust CRN and a memory map which fully describes it, a Turing machine may simulate the CRN by maintaining a tape for each of its species indicating what memory state that species is in. We show that this simulation can be done in real-time for CRNs which use only a finite amount of memory. Although we conjecture that CRNs which use an unbounded amount of memory can also be simulated in real-time (which, if true, would unify the analog and discrete Hartmanis-Stearns Conjectures), finite memory suffices for many real world applications.

The rest of the paper is organized as follows, with many proofs omitted for brevity. Section 2 reviews some necessary preliminaries used in the remainder of the paper. Section 3 resolves the open problem $ALG = \mathbb{R}_{\text{LCRN}} \subsetneq \mathbb{R}_{\text{RTCRN}}$. Section 4 characterizes CRNs in terms of a robust memory map. Lastly, Sect. 5 discusses the consequences of the proceeding sections.

2 Preliminaries

A *Chemical Reaction Network* (CRN), N, is a tuple $N = (S, R)$, where S is a finite number of species and R is a finite set of reactions on those species. In this paper we investigate *deterministic* CRNs, i.e., CRNs under deterministic mass action semantics that are modeled with systems of differential equations [6]. Given a deterministic CRN, let $x_i(t)$ denote the real-valued concentration of X_i at time t for each species $X_i \in S$. Let $\mathbf{x} = (x_1, \ldots, x_n)$ denote the state of N, where $n = |S|$. We write the rate of change of each x_i as $\frac{dx_i}{dt} = f_i(x_1, \ldots, x_n)$ and the rate of change of the entire system as $\frac{d\mathbf{x}}{dt} = \mathbf{f}_N = (f_1, \ldots, f_n)$. Each f_i is a polynomial determined by N [6]. In this paper, rate constants for each reaction in R are integral, and thus each $f_i \in \mathbb{Z}[x_1, \ldots, x_n]$. Furthermore, the initial concentrations of the species, given by an initial state $\mathbf{x}(0) = \mathbf{x}_0$, along with $\mathbf{f}_N$ determine the unique behavior of N. Lastly, when $f_N(\mathbf{z}) = 0$, we call $\mathbf{z}$ a fixed point.

The definition of real-time computable by a CRN used in this paper is given by [11,12]. We repeat the definition here for convenience.

Definition 1. *A real number α is* real-time computable by CRNs *if there exists a CRN $N = (S, R)$ and a species $X \in S$ with the following properties:*

1. *(Integrality.) All rate constants of R are positive integers.*
2. *(Boundedness.) The concentration $x_i(t)$ for each species in S is bounded by a constant β for all time $t \in [0, \infty)$ when $\mathbf{x}_0 = 0$.*
3. *(Real-Time Convergence.) If N is initialized with $\mathbf{x}_0 = 0$, then for all times $t \geq 1$, $||x(t) - |\alpha|| < 2^{-t}$.*

We denote the set of all real-time CRN-computable real numbers as $\mathbb{R}_{RTCRN}$.

Excluding the species that converges to α, the above definition places no restrictions on any species beyond that they be bounded. In many cases, this may be undesirable. The next definition formalizes the notion of converging to a single state, at which point the CRN can be considered finished.

Definition 2. *An* exponentially stable point *of a CRN is a state $\mathbf{z} \in \mathbb{R}_{\geq 0}^n$ for which there exists $\alpha, \delta, C > 0$ such that, if the CRN is initialized to a state $\mathbf{x}_0$ satisfying $|\mathbf{z} - \mathbf{x}_0| < \delta$, then for all times $t \geq 0$, $|\mathbf{z} - \mathbf{x}(t)| \leq Ce^{-\alpha t}|\mathbf{z} - \mathbf{x}_0|$.*

Definition 3. *A real number α is* Lyapunov-CRN computable *if there exists a CRN $N = (S, R)$, a species $X_i \in S$, and a state $\mathbf{z}$ with $\mathbf{z}(X_i) = |\alpha|$ that satisfies the following properties:*

1. *(Integrality.) All rate constants of R are positive integers.*
2. *(Boundedness.) The concentration $x_i(t)$ for each species in S is bounded by a constant β for all time $t \in [0, \infty)$ when $\boldsymbol{x}_0 = 0$.*
3. *(Exponential Stability.) z is an exponentially stable point.*
4. *(Convergence.) If N is initialized with $\boldsymbol{x}_0 = 0$, then $\lim\limits_{t\to\infty} \boldsymbol{x}(t) = \boldsymbol{z}$.*

We denote the set of all Lyapunov-CRN computable real numbers as $\mathbb{R}_{LCRN}$.

Observation 4. *If $\boldsymbol{z}$ is an exponentially stable point of a CRN, then it is a fixed point of that CRN.*

Note that the converse of Observation 4 is not true.

We use *ALG* to denote the set of real algebraic numbers of the rationals. This is the set of real numbers which are the root of some polynomial $f \in \mathbb{Q}[x]$, with rational coefficients.

3 Lyapunov Reals Are Algebraic

To investigate robustness issues in real-time computing, we first look at the relationship between $\mathbb{R}_{LCRN}$ and *ALG* and show that $\mathbb{R}_{LCRN} = ALG$. As a consequence, a bounded CRN may only compute the algebraic numbers reliably in the sense that they exist inside of a potential well. Since Huang et al. proved that $ALG \subsetneq \mathbb{R}_{RTCRN}$ and $ALG \subseteq \mathbb{R}_{LCRN} \subseteq \mathbb{R}_{RTCRN}$ [12], it suffices to show that $ALG = \mathbb{R}_{LCRN}$ to resolve that $\mathbb{R}_{LCRN} \subsetneq \mathbb{R}_{RTCRN}$. We prove this result in two parts. First, we show that every exponentially stable fixed point is isolated. Second, we show that isolated fixed points necessarily have algebraic components.

Let E_N denote the set of exponentially stable points of a CRN, N, and let F_N denote the set of fixed points of N. Recall that fixed points are not necessarily isolated (consider a CRN which does nothing once initialized), however, the set of exponentially stable fixed points, E_N, are isolated in F_N (not just E_N).

Below are two supporting lemmas, as described above. The proofs are omitted for brevity.

Lemma 5. *If $\boldsymbol{z}$ is an exponentially stable point of a CRN, N, then $\boldsymbol{z}$ is isolated in F_N.*

Lemma 6. *If $\boldsymbol{z}$ is a fixed point of a CRN, N, that is isolated in F_N, then the components of $\boldsymbol{z}$ are in ALG.*

Using these lemmas, it is now straightforward to prove the theorem.

Theorem 7. $ALG = \mathbb{R}_{LCRN}$

Proof. Let $\alpha \in \mathbb{R}_{LCRN}$, and let N, X_i, and $\boldsymbol{z}$ be the CRN, designated species, and exponentially stable point that testify to this. By definition, $\boldsymbol{z}$ is exponentially stable; by Lemma 5, $\boldsymbol{z}$ is isolated in F_N; by Lemma 6, every component of $\boldsymbol{z}$ is algebraic. Thus, $\boldsymbol{z}(X_i) = |\alpha|$ is algebraic, and therefore $\alpha \in ALG$.   □

4   A Robust Notion of Memory in CRNs

In the previous section, we concerned ourselves with CRNs which are permitted infinite precision to compute real values robustly in the limit. This excuses several impossibilities for the elegance of its model at the expense of realism. In practice, these CRNs would compute their intended values robustly in approximation and would require only finite time.

In this section we explore the consequences of requiring a CRN to be robust in this sense, that is that they compute values approximately in finite time. In particular, we characterize the behavior of these robust CRNs in terms of these approximations to arrive at a somewhat paradoxical discrete model of analog computing.

Recall that boundedness is one of the three criteria for a real-time CRN. For this section, we use the following definitions of boundedness.

Definition 8. *A CRN* $N = (S, R)$ *is* β*-bounded at* $\boldsymbol{x_0} \in \mathbb{R}^S_{\geq 0}$ *if, when initialized to* $\boldsymbol{x_0}$, *there exists some* $\beta > 0$ *such that* $x < \beta$ *for each* $X \in S$. *Moreover,* N *is uniformly* β*-bounded on* $O \subseteq \mathbb{R}^S_{\geq 0}$ *if there is some* $\beta > 0$ *for which* N *is bounded on each* $\boldsymbol{x_0} \in O$ *by* β.

Unless otherwise specified, a bounded CRN is initialized to the point at which it is bounded. Similarly, a uniformly bounded CRN is initialized to a point at which it is bounded (and is implicitly bounded at any initial point).

There are two natural ways by which a CRN may compute a number α. It may either do so exactly when a species' concentration becomes α or in the limit as per Lyapunov-CRN computability, real-time computability, or some slower manner. Both approaches, however, are imperfect. In the latter case, the concentration of the species computing α either must always maintain a non-zero distance from α after any finite time or, at best, suffers from the same limitation of computing α exactly: the inability to remain at α. The following theorem and corollary formalize this notion.

Theorem 9. *Let* $N = (S, R)$ *be a bounded CRN. For each species* $X \in S$, x *is either constant or the set of times for which* $\frac{dx}{dt} = 0$ *is countable.*

Corollary 10. *Let* $N = (S, R)$ *be a bounded CRN. Pick* $c \in \mathbb{R}_{\geq 0}$. *Then for any non-constant species* $X \in S$, *the set of times* $t \in \mathbb{R}_{\geq 0}$ *where* $x(t) = c$ *is countable.*

It is clear from Corollary 10 that computing an exact value with a CRN is, if not impossible, then a less meaningful concept than one would prefer. This is not inherently problematic as a model of computation. A CRN is capable of computing any computable function in the limit [7].

In each of these models, however, there is the implicit assumption that a CRN may be precisely constructed by which we mean each rate constant and the initial concentration of each species is exactly as prescribed. In practice, this is impractical, which leads us to a notion of robustness. A CRN, informally

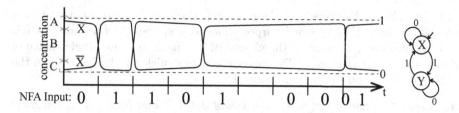

Fig. 1. In [13], Klinge, Lathrop, and Lutz provide a general CRN construction for nondeterministic finite automata (NFAs). These NFAs utilize a dual rail system for each state Z with $z(t) \approx 1$ indicating that the NFA is in state Z at time t and $z(t) \approx 0$ indicating the NFA is not in state Z while the complementary species $\overline{Z}$ has the opposite meaning. Above, we graph the concentration of X and $\overline{X}$ as input changes and show the approximation regions.

speaking, is "robust" if it can tolerate a small perturbation of its concentrations (or rate constants) at any time without affecting its function. This is intuitively a difficult task since changing any such condition clearly alters the solution to the system of ODEs describing the CRN.

Exponentially stable points are a good example of robustness in the following sense. If a CRN manages to get within an ϵ-ball of such a point z, it proceeds to z in the limit without exception. Ideally, a robust CRN would transition from exponentially stable point to exponentially stable point during its computation with some outside force periodically driving it away from each stable equilibrium.

Exponential stability is a far stricter requirement than is necessary to compute a number α, but it does illustrate an important point. If a CRN computes α either in the limit or for longer than a countable set of times, there is always a buffer zone around it which must necessarily be considered in an equivalence class with α. In Fig. 1, this corresponds to the intervals labeled A, B, and C which could be considered equivalence classes for 1, $\frac{1}{2}$, and 0 respectively. We formalize this notion in the following theorems and definitions.

Theorem 11. *Let $N = (S, R)$ be a bounded CRN, and let $X \in S$ be a non-constant species. For any time $t_0 \in \mathbb{R}_{\geq 0}$, there exists a $\delta > 0$ such that for all $t \in (t_0, t_0 + \delta)$, $x(t) \neq x(t_0)$. Moreover, there exists an $\epsilon > 0$ and a $t \in (t_0, t_0 + \delta)$ such that $|x(t) - x(t_0)| > \epsilon$.*

In light of Theorem 11, we state a notion of computation useful (but alone insufficient) for CRNs.

Definition 12. *A CRN $N = (S, R)$ (ϵ, d)-computes a real number $\alpha \in \mathbb{R}_{\geq 0}$ if there is an $X \in S$ and a time $t_0 \in \mathbb{R}_{\geq 0}$ such that $|x(t) - \alpha| < \epsilon$ for all $t \in (t_0, t_0 + d)$.*

Less formally, a CRN (ϵ, d)-*computes* a real number α if it gets close enough to it for a long enough time. To continue our earlier example, the correct choice of ϵ and d make $x(t)$ correctly compute 0 and 1 but never the garbage state $\frac{1}{2}$

in Fig. 1. This underscores that the particular choice of these two parameters is critical for the CRN's intended purpose. Indeed, a species X of a *bounded* CRN so computes every element of the closure of its image for some single choice of d for every ϵ and vice versa! The latter is obvious (pick ϵ to be larger than the CRN's bound), and we formally state the former.

Theorem 13. *Let $N = (S, R)$ be a bounded CRN, and let $\epsilon > 0$. Then there exists a $d > 0$ such that each $X \in S$ (ϵ, d)-computes every element of $cl(x(\mathbb{R}_{\geq 0}))$.*

The following definition resolves this ϵ, d conundrum described above by eliminating any overlap of (ϵ, d)-computed real numbers.

Definition 14. *A CRN $N = (S, R)$ unambiguously computes a set $A \subseteq \mathbb{R}_{\geq 0}$ if for each $\alpha \in A$ there exists a species $X \in S$ which $(\epsilon_\alpha, d_\alpha)$-computes α for some $\epsilon_\alpha, d_\alpha > 0$ and for each distinct $\alpha_1, \alpha_2 \in A$ which X $(\epsilon_{\alpha_1}, d_{\alpha_1})$-computes and $(\epsilon_{\alpha_2}, d_{\alpha_2})$-computes respectively, the intervals $(\alpha_1 - \epsilon_{\alpha_1}, \alpha_1 + \epsilon_{\alpha_1})$ and $(\alpha_2 - \epsilon_{\alpha_2}, \alpha_2 + \epsilon_{\alpha_2})$ are disjoint.*

This notion of unambiguous computation leads directly to a robust notion of CRN memory, but we first state a motivating theorem behind its construction.

Theorem 15. *No CRN can unambiguously compute a somewhere dense subset D of $\mathbb{R}_{\geq 0}$ for any choice of $\epsilon_\alpha, d_\alpha > 0$ for each $\alpha \in D$.*

Theorem 15 shows that many natural encodings of countably infinite sets to bounded intervals cannot be unambiguously computed by a CRN. An example of such is given below where we encode $1 \to .1$, $2 \to .01$, $3 \to .11$, and so on.

Corollary 16. *Let $f : \mathbb{N} \to [0, 1]$ be the map*

$$f(n) = \sum_{i=0}^{\infty} \left(\left\lfloor \frac{n}{2^i} \right\rfloor \mod 2 \right) 2^{-i-1}.$$

No CRN can unambiguously compute the set $f(\mathbb{N})$.

To avoid this problem, any encoding requires an open interval around each value α the CRN must compute wherein the entire interval is considered to be α. Moreover, a CRN can only have countably many such disjoint sets. In our running example, Fig. 1 demonstrates three such intervals for each state species. This leads to the following definition wherein we encode a collection of disjoint open intervals to map to identifying natural numbers.

Definition 17. *Let $c \in \mathbb{R}^+$. A memory map is a map $f : \overline{\mathbb{N}} \to \mathscr{P}([0, c])$ satisfying the following conditions:*

- $f(0) = [0, b)$, where $0 < b \leq c$.
- $\forall n \in \mathbb{Z}^+$, $f(n) = (a, b)$, where $a, b \in \mathbb{Q}$ and $0 < a \leq b \leq c$.
- $\forall m, n \in \mathbb{N}$, if $m \neq n$, then $f(n)$ and $f(m)$ are disjoint.
- $f(\infty) = [0, c] \setminus \bigcup_{n \in \mathbb{N}} f(n)$ and is countable.

. **Definition 18.** *The set of all memory maps on* $[0, c]$ *is* $\mathcal{M}_c$. *The* order *of* $f \in \mathcal{M}_c$, *written* $\mathrm{ord}(f)$, *is the cardinality of the support of* f *over* $\mathbb{N}$.

Definition 19. *The* inverse memory map *of* $f \in \mathcal{M}_c$ *is a map* $f^{\leftarrow} : [0, c] \to \overline{\mathbb{N}}$ *such that for all* $r \in [0, c]$, $r \in f(f^{\leftarrow}(r))$.

In principle, a CRN cannot reasonably be initialized to any state more precise than to an interval of a memory map. Indeed, the consequence of Corollary 10 is the well known fact that if a species ever has a non-zero concentration, it will at almost every time $t > 0$, so no power is gained from being able to initialize a species to 0.

Before proceeding, the definition of a memory map, it should be noted, is *descriptive* of a CRN, not *prescriptive*. Any memory map can model any CRN, but not all memory maps model any particular CRN *well*. For example, any β-bounded CRN can be modeled by the uninteresting memory map that maps every concentration less than β to 0. Similarly, a memory map with randomly chosen intervals is both equally valid and equally ill-suited. We do not yet, however, have all of the definitions necessary to describe what makes for a *good* choice of memory map and so return to this topic later in this section.

Now equipped with a notion of memory, we must define the trajectory of a species X through that memory (and a CRN's trajectory in terms of its species'). This is not inherently clear because a species X must pass over all intermediate memory locations when transitioning between two non-adjacent states. Even if there are only finitely many such intermediary states, including them in the trajectory provides no additional information. That X passes through them during the transition is a direct consequence of x being continuous. In Fig. 1, for example, we never want to include the B interval in our trajectory.

But since each memory state consists of an open interval, X must spend a non-zero length of time inside of it. This brings us back to the definition of (ϵ, d)-computability. If we require X to (ϵ, d)-compute the midpoint of the interval of a memory state with ϵ being half of the interval's width and d being an adjustable parameter, we can arrive at a useful definition of trajectory. To fully formalize this, however, we first have to develop a bit more notation.

Definition 20. *Let* $N = (S, R)$ *be a* β-*bounded CRN. For each* $X \in S$, *let* $f_X \in \mathcal{M}_\beta$ *be a memory map. A species* $X \in S$ *is in the* memory state $m \in \overline{\mathbb{N}}$ *at time* t *if* $x(t) \in f_X(m)$. *Similarly,* N *is in the* memory state $\boldsymbol{m} \in \overline{\mathbb{N}}^S$ *at time* t *if for each* $X \in S$, X *is in the memory state* $\boldsymbol{m}(X)$.

Definition 21. *Let* N *be a* β-*bounded CRN. For* $X \in S$, *let* $f_X \in \mathcal{M}_\beta$ *be a memory map.* X enters *a memory state* $n \in \overline{\mathbb{N}}$ *at time* t_0 *if there exists an* $\epsilon > 0$ *such that for all* $0 < \epsilon' < \epsilon$, $x(t_0 - \epsilon') \notin f_X(n)$ *and* $x(t_0 + \epsilon') \in f_X(n)$. *Similarly,* X leaves n *at time* t_1 *if there exists an* $\epsilon > 0$ *such that for all* $0 < \epsilon' < \epsilon$, $x(t_1 - \epsilon') \in f_X(n)$ *and* $x(t_1 + \epsilon') \notin f_X(n)$. *The* state time *of* X *in* n *(with no intermediate states) is the difference* $t_1 - t_0$.

There are a few consequences to the above definitions worth mentioning. When a species is initialized to a memory state n, it leaves n without first having entered n. A species may also transition from $n \in \mathbb{N}$ to ∞ (or in other words, it may touch the boundary of n) and then return to n, in which case it does not leave or enter n. This is a desirable property as ∞ is not a useful memory state except, perhaps, in the limit as $t \to \infty$. Further, no species may enter or leave the memory state ∞ by definition.

There remain a few edge cases in the above definitions. If a species never entered a memory state n before it leaves n (i.e. it was initialized to n), then we say it entered at time $t = 0$. In a similar vein, if a species never leaves a memory state, we say it leaves at $t = \infty$ purely as a matter of notational convenience (even if in the limit it transitions to the memory state ∞).

Lastly, we remark that sojourn time (arc length) is generally a better measure of runtime for CRNs [3]. In the case of bounded CRNs, however, state time suffices as it is always within a constant factor of sojourn time. This is because each species concentration of a bounded CRN necessarily has a bounded rate of change.

We now have the tools necessary to define the trajectory of a CRN. We first give an informal description here with example and then rigorously define it (see Definition 22). The trajectory of a β-bounded CRN N initialized to $\boldsymbol{x_0}$ is the ordered sequence of memory states obtained as follows. Start from the initial memory state $\boldsymbol{n_0}$. Each time one or more species enters a new memory state for which its state time is at least d, append the new state of N to the sequence. Continue indefinitely or until there are no further memory state changes.

This construction avoids the undesirability of recording in-between memory states of other species as they transition to their next memory state. It also has the added benefit of bringing into the trajectory a notion of a species staying in a memory state for a long time. In a species trajectory, we merely record where the species's concentration goes to but not for how long it stays there. In the memory trajectory of a full CRN, however, if a species's memory state only rarely changes, we can see that behavior in how infrequently it changes in comparison to other species. For example, if one wishes to record a species's memory state at regular intervals, the simple solution is to set up a clock with an appropriate period which has no interaction with the rest of the CRN except to place itself into the memory trajectory as a timestamp.

To complete our running example, Fig. 1 has the following trajectory (for the species X, $\overline{X}$): $(A, C)(C, A)(A, C)(C, A)(A, C)$. In this case, because the construction of the CRN requires that $x + \overline{x} = 1$, a symmetric choice of intervals A and C across $\frac{1}{2}$ causes X and $\overline{X}$ to always be in opposite states at all times.

Using this intuition, we now formally construct the definition of trajectory as follows.

Definition 22. *Let $N = (S, R)$ be a β-bounded CRN at $\boldsymbol{x_0} \in \mathbb{R}_{\geq 0}^{S}$. For $X \in S$, let $f_X \in \mathcal{M}_\beta$ be a memory map. The* memory trajectory *of $\overline{N}$ when N is initialized to $\boldsymbol{x_0}$ with delay $d \in \mathbb{R}^{+}$, written $\boldsymbol{traj}(\boldsymbol{x_0}, d)$, is a sequence of $\mathbb{N}^{S}$ defined by $\boldsymbol{traj}(\boldsymbol{x_0}, d)(n)(X) = m_{\text{last}}(X, \boldsymbol{x_0}, \boldsymbol{m}_{\text{next}}^{n}(\boldsymbol{x_0}, 0, d), d)$ where m_{last} and $\boldsymbol{m}_{\text{next}}^{n}$ are helper functions defined below.*

To define the helper functions in the above definition, let $N = (S,R)$ be a β-bounded CRN at $\boldsymbol{x_0} \in \mathbb{R}^S_{\geq 0}$. For $X \in S$, let $f_X \in \mathcal{M}_\beta$ be a memory map. Let $T(X, \boldsymbol{x_0}, d)$ be the set of times when X enters a memory state for which its state time is at least $d \in \mathbb{R}^+$ when N is initialized to $\boldsymbol{x_0}$, and define $T(\boldsymbol{x_0}, d) = \bigcup_{X \in S} T(X, \boldsymbol{x_0}, d)$.

Next define $\boldsymbol{m}_{\text{next}} : \mathbb{R}^S_{\geq 0} \times \mathbb{R}_{\geq 0} \times \mathbb{R}^+ \to \mathbb{R}_{\geq 0}$ to be the function which, given an initial state $\boldsymbol{x_0}$ of N, a time $t \in \mathbb{R}_{\geq 0}$, and a delay $d > 0$, selects the least $t_0 \in T(\boldsymbol{x_0}, d)$ for which $t_0 > t$.

First, by Corollary 10, $f_X^{\leftarrow} = \infty$ only when X is instantaneously between memory states or if X is constant and initialized to such a value. Both are undesirable system behavior easily avoided. $\boldsymbol{m}_{\text{next}}$ specifically excludes the former from trajectories while the latter is a mere matter of initialization. Unless otherwise specified, we *never* initialize a CRN to such a state even if it is a state for which the CRN is bounded. Second, there is always a least element of each $T(\boldsymbol{x_0}, d)$ for $\boldsymbol{m}_{\text{next}}$ to select since a species X can be in at most two memory states (leaving one for the other) per every d interval of time, which we formally state below.

Lemma 23. *Let $N = (S, R)$ be a β-bounded CRN initialized to $\boldsymbol{x_0} \in \mathbb{R}^S_{\geq 0}$. Fix a delay $d \in \mathbb{R}^+$. Then for any d interval of time, N's memory trajectory contains at most $2|S|$ memory states.*

From this lemma, $\boldsymbol{m}_{\text{next}}$ is well defined. We extend its definition to a recursive form as follows.

$$\boldsymbol{m}^n_{\text{next}}(\boldsymbol{x_0}, t, d) = \begin{cases} \boldsymbol{m}^{n-1}_{\text{next}}(\boldsymbol{x_0}, \boldsymbol{m}_{\text{next}}(\boldsymbol{x_0}, t, d), d) & n > 0 \\ t & n = 0 \end{cases}$$

To finish formalizing the definition of memory trajectory, $m_{\text{last}} : S \times \mathbb{R}^S_{\geq 0} \times \mathbb{R}_{\geq 0} \times \mathbb{R}^+ \to \mathbb{N}$ is the function which, given a species $X \in S$, an initial state $\boldsymbol{x_0}$ of N, a time $t \in \mathbb{R}_{\geq 0}$, and a delay $d > 0$, returns the last memory state $n \in \overline{\mathbb{N}}$ for which species X enters n at a time $t_0 \leq t$ when N is initialized to $\boldsymbol{x_0}$. More intuitively, m_{last} remembers the current memory state of a species while it is transitioning to another memory state.

We can at last now state what makes for a good memory map. The guiding principle behind the choice of memory map is that a CRN in a memory state should behave identically going forward regardless of what particular concentration each species has inside of it. In the spirit of Theorem 15, this then leads to the following natural definition.

Definition 24. *Let $N = (S, R)$ be a uniformly β-bounded CRN and, for each $X \in S$, a memory map $f_X \in \mathcal{M}_\beta$. Fix a delay $d \in \mathbb{R}^+$. N is memory deterministic (with respect to $\{f_X\}_{X \in S}$ and delay d) if there is a function $\delta : \mathbb{N}^S \to \mathbb{N}^S$ such that if $\boldsymbol{m} \in \mathbb{N}^S$ is a memory state in N's memory trajectory, then the next memory state in N's memory trajectory (if one exists) is $\delta(\boldsymbol{m})$. When no such memory state exists, $\delta(\boldsymbol{m}) = \boldsymbol{m}$.*

It is important to note that the transition function δ described in the above definition is relative to which memory state(s) the CRN it describes may be initialized. The behavior of unreachable memory states are outside the scope of the definition. With respect to any such memory state, δ's behavior is unrestricted. In general, δ itself need not necessarily even be computable (although it typically should be). For a well behaved CRN (one which admits many possible initializations), however, δ must satisfy the definition for every valid initialization simultaneously.

Before moving on, observe that *all* bounded CRNs are memory deterministic for *a* choice of memory map. Recall that the memory map which maps all concentrations to 0 is valid for every CRN. Similarly, each CRN modeled with this memory map is memory deterministic with $\delta(\boldsymbol{m}) = \boldsymbol{m}$. Otherwise put, a good choice of memory map for a CRN requires not just that it be memory deterministic but that it is also sufficiently refined to produce a useful model.

Now, unsurprisingly, the notion of a memory map bears a strong resemblance to a Turing machine tape. We know that CRNs and Turing machines are equivalent models [7]. The question remains, however, if one model can outperform the other in some significant way. We can address this question in one direction by providing a means for a Turing machine to simulate a CRN. In general, this is a difficult task. With memory maps, this becomes easier. Since neither model is allowed to speed up indefinitely, we may treat a single step of a Turing machine as a constant length of time. Then we may define that a Turing machine simulates a CRN N if it follows N's memory trajectory on its tape(s). Formally, we have the following definitions.

Definition 25. *Fix $k \in \mathbb{Z}^+$. Let $\Lambda = \{(\boldsymbol{\omega_n}, t_n)\}_{n \in \mathbb{N}}$ be a sequence of tuples in $(\mathbb{Z}_2^*)^k \times \mathbb{R}_{\geq 0}$. A Turing machine with at least k tapes initialized to $\boldsymbol{\omega_0}$ follows Λ if there is a strictly increasing computable sequence $\{s_n\}_{n \in \mathbb{N}}$ of $\mathbb{N}$ such that for each $i \in \mathbb{Z}_k$, the contents of tape T_i at step s_n is $\boldsymbol{\omega_n}(i)$. Similarly, M real-time follows Λ if there is a constant $c > 0$ such that each $s_n \leq ct_n$*

Definition 26. *Let $N = (S, R)$ be a (uniformly) β-bounded CRN and, for each $X \in S$, let $f_X \in \mathcal{M}_\beta$. Fix a delay $d \in \mathbb{R}^+$. A Turing machine M follows N according to $\{f_X\}_{X \in S}$ with delay d if for each $X \in S$ there exists a computable injective map $itoa_X : \mathbb{N} \to \mathbb{Z}_2^*$ such that M follows*

$$\Lambda = \{(\boldsymbol{itoa}(\boldsymbol{traj}(\boldsymbol{x_0}, d)(n)), \boldsymbol{m}_{\text{next}}^n(\boldsymbol{x_0}, 0, d))\}_{n \in \mathbb{N}}$$

when initialized to $\boldsymbol{x_0} \in \mathbb{R}_{\geq 0}^S$ (for every initialization $\boldsymbol{x_0} \in \mathbb{R}_{\geq 0}^S$ for which N is bounded), where $\boldsymbol{itoa}(\boldsymbol{traj}(\boldsymbol{x_0}, d)(n))(X) = itoa_X(\boldsymbol{traj}(\boldsymbol{x_0}, d)(n)(X))$ for $X \in S$ and $n \in \mathbb{N}$. Similarly, M real-time follows N according to $\{f_X\}_{X \in S}$ with delay d if M real-time follows Λ when initialized to $\boldsymbol{x_0} \in \mathbb{R}_{\geq 0}^S$ (for every initialization $\boldsymbol{x_0} \in \mathbb{R}_{\geq 0}^S$ for which N is bounded).

We can extend our running example to these definitions as follows. For the itoa functions, interval A maps to 1, B maps to 10, and C maps to 0. Moreover, since the CRN was constructed directly from a finite automaton, it only takes two steps to compute each subsequent memory state and write it to the appropriate tape. If follows trivially, then, that there is a Turing machine which real-time follows the CRN.

More generally, since Turing machines and CRNs are equivalent models [7], there is always a Turing machine that follows any CRN N. The more interesting (and far more difficult) question is if there always exists a Turing machine M and some choice of itoa functions for which M real-time follows N. Intuitively, analog computing *should* be more efficient than discrete computing in some respect. Indeed, were a CRN either unbounded or if it were allowed an unbounded number of species, this is easy to show. To see why this is less certain for robust, bounded CRNs, we need a few lemmas.

The natural first question to ask is how can a Turing machine can keep up in real-time with a CRN from the definition of real-time following. A CRN, after all, is allowed to change all of its species concentrations simultaneously while a Turing machine, following the CRN's memory trajectory and not directly simulating the CRN, must keep all but one (except in the unlikely case where two or more species change memory state at the exact same time) of its species-tracking tapes effectively constant between memory trajectory transitions.

This is not a limitation since a species must linger in a memory state for a minimum length of time. A CRN with n species and delay d can only experience at most n memory states in every open interval of length d (see Lemma 23). This is what makes a Turing machine M real-time following a CRN occur in real-time. It follows that M can compute each of these state changes sequentially while only requiring a constant factor of $|S|$ more time in the worst case.

The remaining difficulty is to show that a bounded CRN cannot 'cheat' in the sense that a Turing machine would require an infinite alphabet or an infinite number of states or tapes to real-time follow it. We show this is the case when each memory map has only finite order and the transitions between memory states is memory deterministic.

Theorem 27. *Let $N = (S, R)$ be a uniformly β-bounded CRN, let $f_X \in \mathcal{M}_\beta$ for each $X \in S$ with $\mathrm{ord}(f_X) < \infty$, and let $d \in \mathbb{R}^+$. If N is memory deterministic, then there is a Turing machine which real-time follows N according to $\{f_X\}_{X \in S}$ with delay d.*

Corollary 28. *Let $N = (S, R)$ be a uniformly β-bounded CRN, and let $f_X \in \mathcal{M}_\beta$ for each $X \in S$. Fix a delay $d \in \mathbb{R}^+$. If N is memory deterministic and N's memory trajectory is either finite or there exists a memory state which appears at least twice in it, then there is a Turing machine which real-time follows N according to $\{f_X\}_{X \in S}$ with delay d.*

5   Discussion

In this paper, we have shown that *only* the algebraic real numbers are computable by CRNs using exponentially stable equilibria. Intuitively, this means that every transcendental real number cannot be computed robustly by a CRN in the sense of Definition 3. This led us to explore in Sect. 4 what it means for a CRN to compute robustly. We started from two notions of computation. First, a CRN can compute a value exactly, which a non-constant CRN can achieve only for a measure zero length of time. Second, a CRN may compute a value in the limit, which has two problems of its own. A CRN never precisely achieves a value computed in the limit. Moreover, we showed earlier in Theorem 7 that only the algebraic numbers can be so computed reliably. Any non-algebraic number, if the CRN is improperly initialized with any epsilon error, cannot be computed in the limit.

These limitations led us to ask what happens when we require a CRN to behave identically for a range of inputs rather than a single set of concentrations. The result was the notion of a memory map, a strangely discrete model of an analog implementation of computation. Arguably, under this model, a CRN's reactions correspond to transitions between states of a Turing machine while species concentrations correspond to tape states.

This ultimately led to Theorem 27. In the more familiar terminology of the discrete world, it tells us that a robust CRN is no more capable of executing a NFA than a Turing machine is. This is perhaps unsurprising. The main advantage a CRN has to leverage over a Turing machine is in its ability to rewrite its entire tape with a new word of any length. With finite memory, this advantage is lost.

Notice, however, that Theorem 27 says nothing about the *existence* of a robust CRN capable of simulating a NFA. For that, we turn to [13] for a CRN with a more restrictive notion of robustness which nonetheless satisfies the definitions we derived here and Theorem 27. We briefly summarize this below as an illustrative example.

Given a NFA M, a CRN N is constructed with two species, X_q and $\overline{X}_q$, for each state q of M. These species alternatively take concentrations close to 1 or 0 to represent M being in state q or not in state q respectively for X_q and vice versa for $\overline{X}_q$. The appropriate memory map for each of these species would be to map 0 to an interval around 0, 1 to an interval around 1, and 2 to everything in-between. The correct delay to choose for this CRN is the length of the clock cycle (which also admits an identical memory map). For the input signal (which, again, admits an identical memory map), we may assume that there is an external CRN generating it which makes the N memory deterministic.

First, note that N is uniformly bounded on all of its valid inputs. Moreover, if we apply Theorem 27 to the CRN described above, we obtain a Turing machine which not only behaves identically but can be transformed back into the same CRN [13]. As such, these are truly inverse statements. Moreover, the theorem can be applied to more general cases as well.

Given a Turing machine M, Fages et al. construct a GPAC-generable function (easily translated into the CRN world) that simulates M within bounded time and tape space [7]. The input parameters for each bound can be adjusted, of course, but once fixed, the resulting simulation permits a single memory map model for all of its input configurations to which Corollary 28 applies. In a sense, these, too, are inverse statements.

Now the question becomes where to go from here. It is known that, given an NFA, there is a robust CRN which simulates it in real-time [13]. Similarly, we have provided a proof that, given a robust CRN with memory maps of only finite order, there is a Turing machine which real-time follows it. In short, for the regular languages, robust CRNs and Turing machines are fully equivalent models with neither having an advantage over the other. We conjecture that the same is true of an arbitrary robust CRN, that is given a robust CRN with a memory deterministic collection of memory maps, there is a Turing machine which real-time follows it. This, if true, has several important implications.

First, it's known that CRNs and Turing machines can simulate each other with a polynomial-time slowdown [3]. If this conjecture is true, even in a more restricted form, it would eliminate the slowdown from a Turing machine simulating a CRN.

Of perhaps more interest is the Hartmanis-Stearns Conjecture (HSC) [10]. Both Turing machines and robust CRNs are clearly capable of outputting the digits of a rational number in real-time. For Turing machines, this means writing to some output tape. For CRNs, this *does not* mean outputting a concentration but rather raising a concentration high or low in a memory trajectory in the appropriate sequence. In this manner, assuming our stated conjecture, then if one could construct a robust CRN to output the digits of a nonrational algebraic number, it would also resolve the HSC for Turing machines.

Acknowledgments. The authors thank anonymous reviewers for useful feedback and suggestions. This research supported in part by NSF grants 1900716 and 1545028.

References

1. Bournez, O., Campagnolo, M.L., Graça, D.S., Hainry, E.: The general purpose analog computer and computable analysis are two equivalent paradigms of analog computation. In: Cai, J.-Y., Cooper, S.B., Li, A. (eds.) TAMC 2006. LNCS, vol. 3959, pp. 631–643. Springer, Heidelberg (2006). https://doi.org/10.1007/11750321_60
2. Bournez, O., Fraigniaud, P., Koegler, X.: Computing with large populations using interactions. In: Rovan, B., Sassone, V., Widmayer, P. (eds.) MFCS 2012. LNCS, vol. 7464, pp. 234–246. Springer, Heidelberg (2012). https://doi.org/10.1007/978-3-642-32589-2_23
3. Bournez, O., Graça, D.S., Pouly, A.: Polynomial time corresponds to solutions of polynomial ordinary differential equations of polynomial length. J. ACM **64**(6) (2017). https://doi.org/10.1145/3127496
4. Cappelletti, D., Ortiz-Muñoz, A., Anderson, D.F., Winfree, E.: Stochastic chemical reaction networks for robustly approximating arbitrary probability distributions. Theoret. Comput. Sci. **801**, 64–95 (2020). https://doi.org/10.1016/j.tcs.2019.08.013
5. Clamons, S., Qian, L., Winfree, E.: Programming and simulating chemical reaction networks on a surface. J. R. Soc. Interface **17**(166), 20190790 (2020). https://doi.org/10.1098/rsif.2019.0790
6. Epstein, I.R., Pojman, J.A.: An Introduction to Nonlinear Chemical Dynamics: Oscillations, Patterns, and Chaos. Oxford University Press, Waves (1998)

7. Fages, F., Le Guludec, G., Bournez, O., Pouly, A.: Strong turing completeness of continuous chemical reaction networks and compilation of mixed analog-digital programs. In: Feret, J., Koeppl, H. (eds.) CMSB 2017. LNCS, vol. 10545, pp. 108–127. Springer, Cham (2017). https://doi.org/10.1007/978-3-319-67471-1_7

8. Furcy, D., Summers, S.M., Wendlandt, C.: Self-assembly of and optimal encoding within thin rectangles at temperature-1 in 3D. Theoret. Comput. Sci. (2021). https://doi.org/10.1016/j.tcs.2021.02.001

9. Hader, D., Patitz, M.J.: Geometric tiles and powers and limitations of geometric hindrance in self-assembly. Nat. Comput. **20**(2), 243–258 (2021). https://doi.org/10.1007/s11047-021-09846-2

10. Hartmanis, J., Stearns, R.E.: On the computational complexity of algorithms. Trans. Am. Math. Soc. **117**, 285–306 (1965). http://www.jstor.org/stable/1994208

11. Huang, X., Klinge, T.H., Lathrop, J.I.: Real-time equivalence of chemical reaction networks and analog computers. In: Thachuk, C., Liu, Y. (eds.) DNA 2019. LNCS, vol. 11648, pp. 37–53. Springer, Cham (2019). https://doi.org/10.1007/978-3-030-26807-7_3

12. Huang, X., Klinge, T.H., Lathrop, J.I., Li, X., Lutz, J.H.: Real-time computability of real numbers by chemical reaction networks. Nat. Comput. **18**(1), 63–73 (2018). https://doi.org/10.1007/s11047-018-9706-x

13. Klinge, T.H., Lathrop, J.I., Lutz, J.H.: Robust biomolecular finite automata. Theoret. Comput. Sci. **816**, 114–143 (2020). https://doi.org/10.1016/j.tcs.2020.01.008

14. Klinge, T.H., Lathrop, J.I., Moreno, S., Potter, H.D., Raman, N.K., Riley, M.R.: ALCH: an imperative language for chemical reaction network-controlled tile assembly. In: Geary, C., Patitz, M.J. (eds.) 26th International Conference on DNA Computing and Molecular Programming (DNA 26). Leibniz International Proceedings in Informatics (LIPIcs), vol. 174, pp. 6:1–6:22. Schloss Dagstuhl-Leibniz-Zentrum für Informatik, Dagstuhl (2020). https://doi.org/10.4230/LIPIcs.DNA.2020.6

15. Severson, E.E., Haley, D., Doty, D.: Composable computation in discrete chemical reaction networks. Distrib. Comput. (1), 1–25 (2020). https://doi.org/10.1007/s00446-020-00378-z

16. Turing, A.M.: On computable numbers, with an application to the Entscheidungs problem. Proc. Lond. Math. Society **s2–42**(1), 230–265 (1937). https://doi.org/10.1112/plms/s2-42.1.230

Zero-Knowledge Proof Protocol
for Cryptarithmetic Using Dihedral Cards

Raimu Isuzugawa[1]([✉]) [ID], Daiki Miyahara[1,2] [ID], and Takaaki Mizuki[1,2]([✉]) [ID]

[1] Tohoku University, Sendai, Japan
raimu.isuzugawa.q6@dc.tohoku.ac.jp, mizuki+lncs@tohoku.ac.jp
[2] National Institute of Advanced Industrial Science and Technology (AIST),
Tokyo, Japan

Abstract. Cryptarithmetic, also known as Verbal Arithmetic or Word
Addition, is a popular pencil puzzle in which the aim is to deduce which
letter corresponds to which numeral, given a mathematical equation in
which each numeral (from 0 to 9) has been replaced with a unique letter.
The most famous instance of this puzzle is probably "SEND + MORE =
MONEY", whose solution is "9567 + 1085 = 10652", i.e., S = 9, E = 5,
N = 6, D = 7, M = 1, O = 0, R = 8, and Y = 2. In this study, we construct
a physical zero-knowledge proof protocol for a Cryptarithmetic puzzle:
That is, our protocol enables a prover who knows a solution to the puzzle
to convince a verifier that he/she knows the solution without revealing
any information about it. The proposed protocol uses a physical deck of
"dihedral cards," which were developed by Shinagawa in 2019.

Keywords: Cryptarithmetic · Dihedral cards · Physical
zero-knowledge proof · Card-based cryptography

1 Introduction

Cryptarithmetic, also known as Verbal Arithmetic or Word Addition, is a
famous pencil puzzle: given an equation, such as "SEND + MORE = MONEY"
(Fig. 1(a)), where each numeral from 0 to 9 has been replaced with a unique
letter, one has to guess which letter corresponds to which numeral. The solution
to the aforementioned example is presented in Fig. 1(b): That is, the correspon-
dences are S $\mapsto$ 9, E $\mapsto$ 5, N $\mapsto$ 6, D $\mapsto$ 7, M $\mapsto$ 1, O $\mapsto$ 0, R $\mapsto$ 8, and
Y $\mapsto$ 2.

The rules for solving Cryptarithmetic puzzles are as follows.

1. Any letter (at every position) corresponds to the same numeral, and different
   letters correspond to different numerals.
2. The most significant digit (letter) must not correspond to 0.
3. After all letters are replaced with their numerals, the resulting equation must
   be mathematically correct.

In this paper, we shall construct a *zero-knowledge proof* protocol for
Cryptarithmetic. We begin by explaining what zero-knowledge proof protocols
for pencil puzzles are.

© Springer Nature Switzerland AG 2021
I. Kostitsyna and P. Orponen (Eds.): UCNC 2021, LNCS 12984, pp. 51–67, 2021.
https://doi.org/10.1007/978-3-030-87993-8_4

```
    S E N D              9 5 6 7
  + M O R E            + 1 0 8 5
  ─────────            ─────────
  M O N E Y            1 0 6 5 2
      (a)                  (b)
```

Fig. 1. A puzzle instance of Cryptarithmetic and its solution

1.1 Zero-Knowledge Proofs for Puzzles

Consider a situation in which there are two players: a prover P and a verifier V; the prover P knows a solution w to an instance x of a puzzle (such as Cryptarithmetic and Sudoku[1]), whereas the verifier V does not know any solution to x. The verifier V spent considerable time attempting to find a solution to x, but V was unable to find it. Thus, V becomes skeptical and asks P to prove that there is a solution to x. However, if P only shows the solution w to V, the puzzle instance x will not be worth solving. A *zero-knowledge proof* protocol, whose concept was first conceived in [6], can solve this dilemma: It enables P to convince V of the existence of w without revealing any information about w (that only P knows), satisfying the following three properties.

Completeness. If P knows w, then V is convinced of the existence of w.
Extractability. If P does not know w, then V is not convinced.
Zero-knowledge. V does not obtain any information about w.

In 2007, Gradwohl et al. [7,8] first constructed zero-knowledge proof protocols for Sudoku using physical daily-use objects such as a deck of playing cards. Since then, many zero-knowledge proof protocols for pencil puzzles using a deck of physical cards, which we call *card-based ZKP protocols*, have been proposed, such as those for Akari [1], Hashiwokakero [27], Hitori [21], Juosan [14], Kakuro [1,15], KenKen [1], Makaro [2], Masyu [10], Nonogram [3,22], Norinori [4], Numberlink [24,25], Nurikabe [21], Ripple Effect [26], Slitherlink [10,11], Sudoku [23,28,29], and Takuzu [1,14]. These physical zero-knowledge proof protocols do not depend on computers or programs; hence, it is relatively easy for lay people to perform zero-knowledge proof and/or to understand its concept.

1.2 Our Contribution

It should be noted that all the pencil puzzles listed above are played with a rectangular grid (consisting of many cells): That is, all the existing card-based ZKP protocols have been designed to manipulate a grid with numbers and/or symbols. By contrast, Cryptarithmetic, for which this study shall design a zero-knowledge proof protocol, is played not with a grid but with an equation, as already seen in Fig. 1. Therefore, another technique or treatment is required to

---

[1] Sudoku is the most famous pencil puzzle, which has been published by NIKOLI Co., Ltd. (https://www.nikoli.co.jp/en/).

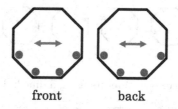

front   back

Fig. 2. Dihedral card of $2m$-sided polygon for $m = 4$

construct a card-based ZKP protocol for Cryptarithmetic. Furthermore, while most of the existing protocols use a deck of physical cards consisting of black and red cards ♣ ♣ ⋯ ♡ ♡ ⋯ and/or number cards 1 2 3 ⋯, these cards are not suitable for computing an arithmetic addition; in particular, computing a carry causes a heavy load (e.g., [17]). Therefore, we need to consider other types of physical cards.

In this paper, we construct a zero-knowledge proof protocol for Cryptarithmetic using *dihedral cards*, as illustrated in Fig. 2; these novel cards were proposed by Shinagawa in 2019 [30,31][2]. That is, using our proposed protocol, a prover P who knows the solution to a given Cryptarithmetic puzzle can convince a verifier V that P knows the solution without revealing any information about it. As will be seen in Sects. 2 and 3, the dihedral cards are suitable for Cryptarithmetic because they can efficiently compute an arithmetic addition with a carry (compared with other regular polygon cards [32], or a normal deck of cards as mentioned above). After we present our protocol in Sect. 3, we evaluate its performance and demonstrate its correctness in Sect. 4. The paper is concluded in Sect. 5.

Although Cryptarithmetic puzzles can have multiple solutions or can be an equation whose left-hand side has more than two terms (for addition) [33–35], we focus on Cryptarithmetic puzzles with the addition of exactly two terms such that there is a unique solution (as shown in Fig. 1) throughout this paper. (We will revisit this point in Sect. 5.)

Note that if we allow the base, denoted by k, in arithmetic to be arbitrary (aside from $k = 10$, i.e., decimal arithmetic), the decision problem for Cryptarithmetic becomes NP-complete [5] (where k is not fixed). The computational complexity of Cryptarithmetic has been studied, including methods for efficiently deriving solutions using genetic algorithm [13] and automata that accept Cryptarithmetic problems for bases $k \leq 7$ [19]. If we fix k, say $k = 10$ as in this paper, then we can find a solution to a given Cryptarithmetic problem (if any) by enumerating at most 10! assignments of numerals to the given letters. However, pencil puzzles such as Cryptarithmetic are usually solved with a pen and sheets of paper; hence, without the aid of a computer, 10! possibilities cannot be enumerated by hands; thus, it is worthwhile to perform zero-knowledge proofs even for puzzles in P.

---

[2] Figures 2 to 11 were created based on the figures presented in [31].

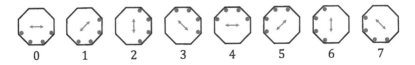

Fig. 3. How to represent integers using a dihedral card of $m = 4$

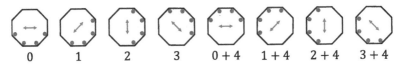

Fig. 4. How to represent integers using a dihedral card considering a carry

2   Preliminaries: Dihedral Cards [30, 31]

In this section, we describe the dihedral cards proposed by Shinagawa [30,31], operations on them, and some sub-protocols that will be used when constructing our protocol in Sect. 3.

2.1   Dihedral Cards

A *dihedral card* is a regular polygon card marked with *invisible ink*. We use dihedral cards, each of which is a regular $2m$-sided polygon, to conduct computations on a finite field $\mathbb{Z}/\mathbb{Z}_m = \{0, 1, \ldots, m-1\}$ for a positive integer m. Figure 2 shows an example of a dihedral card of a $2m$-sided polygon for $m = 4$, i.e., a regular octagon card. The blue arrows and dots in Fig. 2 have been drawn using invisible ink. Specifically, a symmetric bidirectional arrow is marked in the center, and m dots corresponding to m consecutive vertices (among $2m$ ones) are marked. The same pattern is drawn on the back of the card, i.e., every vertex marked with a dot in the front also has a dot in the back.

The physical property of invisible ink guarantees that blue arrows and dots are invisible to the naked eye, but they can be made visible by illuminating with black light. Therefore, markings on a dihedral card can only be confirmed when illuminated with black light.

Because dihedral cards have the shape of a regular $2m$-sided polygon, we can simply represent every integer from 0 to $2m - 1$ depending on the angle at which the cards are placed, as shown in Fig. 3. However, in our proposed protocol, the values from m to $2m - 1$ are treated as the values from 0 to $m - 1$ with a carry, as shown in Fig. 4. Given a $2m$-sided polygon card having a value $x \in \mathbb{Z}/\mathbb{Z}_{2m}$, if we rotate it by $c\pi/m$ degrees for an integer c, its value is changed from x to $x + c \bmod 2m$; we refer to this action as "rotating a card by a degree c."

In addition to rotation, a dihedral card can be transformed by flipping it face up or down based on a certain axis of rotation. In this study, we use three axes, i.e., three flipping methods: Flip the card vertically, as shown in Fig. 5; diagonally, as shown in Fig. 6; and horizontally, as shown in Fig. 7.

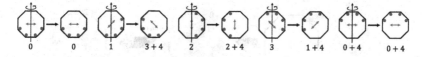

Fig. 5. Flipping a card vertically

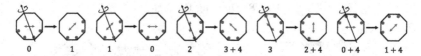

Fig. 6. Flipping a card diagonally

Remember that the value of a dihedral card can be revealed by illuminating with black light. Next, we describe methods for partially revealing the values of a dihedral card. That is, assume that, given a $2m$-sided polygon card with a value $x \in \mathbb{Z}/\mathbb{Z}_{2m}$, we want to obtain the one-bit value $p(x \geq m)$ and/or the m-valued $x \bmod m$. Here, we define $p(x \geq m) = 1$ if $x \geq m$, and $p(x \geq m) = 0$ if $x < m$. We call $p(x \geq m)$ the *sign* of the card, and $x \bmod m$ the mod-m *value* of the card. This can be achieved by covering areas that are not related to the value to be observed; Fig. 8 and Fig. 9 illustrate how to reveal the sign and the mod-m value of a card, respectively, using covers of special shapes. As can be verified in Fig. 4, the sign of a card determines whether or not the digit has a carry, and a dot mark at the point where the black light is irradiated, as shown in Fig. 8, corresponds to this sign. On the other hand, the mod-m value can be obtained by looking only at the arrow in the center illuminated by the black light because its direction reveals the mod-m value only (without its possible carry). In Fig. 8 and Fig. 9, the sign $p(x \geq m)$ is 0, and the mod-m value is 1.

2.2 Shuffle Operations on Dihedral Cards

Here, considering a sequence of dihedral cards, we describe two shuffle operations. Let a positive integer m be fixed, and denote by $[\![x]\!]$ a $2m$-sided polygon card with a value of $x \in \mathbb{Z}/\mathbb{Z}_{2m}$. For a positive integer i, we define $[i] = \{1, 2, \ldots, i\}$. Given a sequence of ℓ regular $2m$-sided dihedral cards $([\![x_1]\!], [\![x_2]\!], \ldots, [\![x_\ell]\!])$, we consider two types of shuffle, as follows.

Rotation Shuffle. A *rotation shuffle* with a set $T \subseteq [\ell]$ rotates all cards whose positions are in T by a uniformly distributed random number $r \in \mathbb{Z}/\mathbb{Z}_{2m}$; that is, it shuffles all cards specified by T together. In this case, the

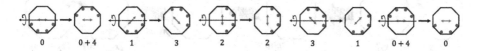

Fig. 7. Flipping a card horizontally

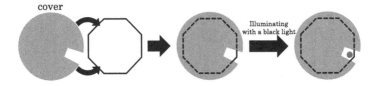

Fig. 8. Opening the sign

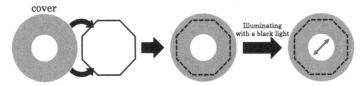

Fig. 9. Opening the mod-m value

sequence of the cards $(\llbracket x_1 \rrbracket, \llbracket x_2 \rrbracket, \ldots, \llbracket x_\ell \rrbracket)$ becomes $x_i \to x_i + r \pmod{2m}$ when $i \in T$, and $x_i \to x_i$ when $i \notin T$.

Two-sided Rotation Shuffle. A *two-sided rotation shuffle* is an operation that randomly rotates all cards whose positions are in $T \subseteq [\ell]$ by π. In this case, the card sequence $(\llbracket x_1 \rrbracket, \llbracket x_2 \rrbracket, \ldots, \llbracket x_\ell \rrbracket)$ becomes $x_i \to x_i + rm \pmod{2m}$ when $i \in T$, and $x_i \to x_i$ when $i \notin T$, where r is a random bit $r \in \{0, 1\}$. This shuffle can be implemented using two clips to fix and rotate the cards, as illustrated in Fig. 10.

2.3 Protocols with Dihedral Cards

In this subsection, we briefly introduce several basic protocols for computations working on dihedral cards [30, 31]; refer to [31] for details.

2.3.1 Initialization Protocol

The initialization protocol takes as input a card $\llbracket x \rrbracket$ such that $x \in \mathbb{Z}/\mathbb{Z}_{2m}$ and initializes its value to 0: $\llbracket x \rrbracket \Rightarrow \llbracket 0 \rrbracket$. It proceeds as follows.

1. Apply a rotation shuffle to the card.
2. Illuminate the whole card with black light and let the opened value be $x' \in \mathbb{Z}/\mathbb{Z}_{2m}$.
3. Rotate the card by a degree $-x'$.

This protocol requires one shuffle.

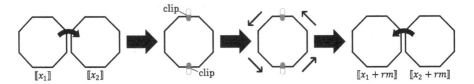

Fig. 10. Implementation of two-sided rotation shuffle ($\ell = 2$)

2.3.2   Addition Protocol

The addition protocol takes as input two cards $[\![x_1]\!], [\![x_2]\!]$ such that $x_1, x_2 \in \mathbb{Z}/\mathbb{Z}_{2m}$, and outputs the arithmetic addition of the two (and $[\![0]\!]$):

$$([\![x_1]\!], [\![x_2]\!]) \Rightarrow ([\![0]\!], [\![x_1 + x_2 \bmod 2m]\!]).$$

It proceeds as follows.

1. Flip the left card vertically, as shown in Fig. 5, to obtain $[\![-x_1]\!]$.
2. Apply a rotation shuffle to the sequence of two cards.
3. Illuminate the entire left card with black light. Let the opened value be $x' \in \mathbb{Z}/\mathbb{Z}_{2m}$.
4. Rotate the sequence of cards by a degree $-x'$.

This protocol also requires one shuffle.

2.3.3   Sign Normalization Protocol

The sign normalization protocol takes as input a card $[\![x]\!]$ such that $x \in \mathbb{Z}/\mathbb{Z}_{2m}$ and changes its value to $x \bmod m$:

$$[\![x]\!] \Rightarrow [\![x \bmod m]\!].$$

It proceeds as follows.

1. Apply a two-sided rotation shuffle to the card.
2. Reveal the sign of the card (using the method illustrated in Fig. 8). Let $s' \in \{0, 1\}$ be the sign of the card.
3. Rotate the card by a degree $s'm$.

This protocol uses one shuffle.

2.3.4   Sign-to-Value Protocol

The sign-to-value protocol takes as input a card $[\![x]\!]$ such that $x \in \mathbb{Z}/\mathbb{Z}_{2m}$ (along with card $[\![0]\!]$), and outputs the sign of the card (and $[\![0]\!]$):

$$([\![x]\!], [\![0]\!]) \Rightarrow ([\![p(x \geq m)]\!], [\![0]\!]).$$

It proceeds as follows.

1. Apply a two-sided rotation shuffle to the sequence of two cards.
2. Reveal the sign of the left card. Let $s_1 \in \{0, 1\}$ be the revealed sign.
3. Rotate the right card by a degree $s_1 m$.
4. Apply the initialization protocol to the left card. We now have $([\![0]\!], [\![p(x \geq m) \cdot m]\!])$.

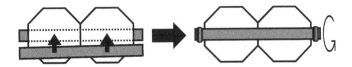

Fig. 11. Implementation of uniform random flipping

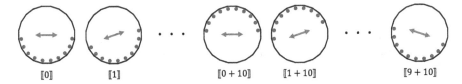

$[\![0]\!]$ $[\![1]\!]$ $[\![0+10]\!]$ $[\![1+10]\!]$ $[\![9+10]\!]$

Fig. 12. Dihedral cards of regular 20-sided polygons

5. Consider a diagonal axis (as in Fig. 6) for the left card and a horizontal axis (as shown in Fig. 7) for the right card. Then, the cards are randomly flipped together based on these axes; to achieve this, after adjusting the degree of the left card, fix the two cards together with two plates, as illustrated in Fig. 11, and repeatedly rotate them quickly.
6. Reveal the sign of the right card. Let $s_2 \in \{0, 1\}$ be the revealed sign.
   (a) If $s_2 = 0$, output the current sequence.
   (b) If $s_2 = 1$, rotate the right card by a degree m, and then, flip the left card diagonally (as shown in Fig. 6).

Three shuffles are required for the sign-to-value protocol.

3   Zero-Knowledge Proof Protocol for Cryptarithmetic

In this section, we construct a card-based zero-knowledge proof protocol for Cryptarithmetic using dihedral cards, utilizing the basic sub-protocols introduced in the previous section.

First, in Sect. 3.1, we propose a copy protocol for use in our proposed protocol. This copy protocol creates multiple dihedral cards with the same value from one dihedral card without revealing any information about its value. We then describe the procedure for our proposed protocol in Sect. 3.2. To deal with decimal arithmetic in Cryptarithmetic, we set $m = 10$, i.e., our protocol uses dihedral cards, each of whose shape is a regular 20-sided polygon, as illustrated in Fig. 12.

In Sect. 4, we will evaluate the numbers of cards and shuffles required for executing our proposed protocol and prove that our protocol satisfies the three properties of the zero-knowledge proof.

3.1   How to Duplicate Commitment

Let us call a dihedral card with a value $x \in \mathbb{Z}/\mathbb{Z}_{2m}$ a *commitment* to x. We present a copy protocol that duplicates a given commitment. As can be observed in Sect. 3.2, we use this copy protocol to duplicate a commitment to every i, $0 \leq i \leq 9$, when setting up our proposed protocol.

Given a commitment $[\![x]\!]$, our copy protocol making $\ell\,(\geq 2)$ copied commitments proceeds as follows.

1. Place a sequence of ℓ dihedral cards $[\![0]\!]$, all having a value of 0, next to the given commitment to be copied.
2. Apply the addition protocol to the sequence of cards so that the value of the given commitment is added to all the ℓ cards, resulting in a sequence of ℓ commitments ($[\![x]\!], \ldots, [\![x]\!]$). (Note that the addition protocol presented in Sect. 2.3.2 takes only two cards as input, but one can easily extend it by rotating the $\ell + 1$ cards together.)

In this protocol, the given commitment is duplicated by adding its value to the desired number of dihedral cards that we want to obtain. Thus, it requires ℓ dihedral cards as well as a given commitment, and requires only one shuffle.

3.2   Procedure

In this subsection, we describe the procedure for our proposed protocol. Given a Cryptarithmetic problem, our protocol enables a prover P who knows the solution to the problem to convince a verifier V that P knows the solution without revealing any information about the solution. It consists of four phases: Setup, Adding least significant digits (half adder), Adding higher digits (full adder), and Verification.

Setup. In this phase, dihedral cards corresponding to the solution are created.

1. Prepare a commitment to i for every i, $0 \leq i \leq 9$, i.e., $[\![0]\!], \ldots, [\![9]\!]$. The values of the commitments should be disclosed so that V can be convinced that every commitment corresponds to a distinct integer.
2. Prepare *symbolic cards* corresponding to the letters appearing in the puzzle instance, as illustrated in Fig. 13; this example corresponds to the puzzle shown in Fig. 1, i.e., we have eight cards with a letter S, E, N, D, M, O, R, or Y on their front, along with two *dummy* cards with blank surfaces, where all 10 cards have indistinguishable backs. The letters on the front can be numbers (as in the case of playing cards), but for the sake of clarity, we use symbolic cards that have the same letters that appeared in the puzzle instance. If the number of letters appearing in the puzzle is less than 10, dummy cards are used for the missing letters, as in the example above.
3. Remember that only the prover P knows the solution, i.e., the one-to-one correspondence between numerals and letters. The prover P takes all the symbolic cards in his/her hand, and places them face-down below the 10 commitments (which were prepared in the first step), such that each pair of a commitment and a symbolic card follows the one-to-one correspondence, as illustrated in Fig. 14, where a dummy card is placed if there is no letter

corresponding to that numeral. Because commitments to 3 and 4 do not appear in the example solution, dummy cards with blank surfaces are placed below them. Note that the face-down 10 symbolic cards have been placed secretly by P without V knowing their order.

Fig. 13. Examples of symbolic cards

4. Fix the 10 pairs of commitments and symbolic cards (or dummy cards) using envelopes or clips and then shuffle them by hand. This shuffling operation is called a *pile-scramble shuffle* [9]. The shuffle is performed by P and/or V; they can repeat shuffling until both of them are satisfied.
5. Turn over all the symbolic cards to see the mapping from letters to commitments. If the revealed card is a dummy card, the corresponding commitment can be discarded after applying a rotation shuffle.
6. Remember that a puzzle instance has an equation where the left-hand side is an addition of two sequences of letters; without loss of generality, we assume that the number of letters in the second term on the left-hand side is greater than or equal to that in the first term. Repeatedly execute the copy protocol presented in Sect. 3.1 to ensure that we have a sufficient number of duplicated commitments to accomplish the following: (i) for every letter in the first term on the left-hand side of addition (in the puzzle instance), place one commitment corresponding to that letter; (ii) for every letter in the second term, place two commitments corresponding to that letter; and (iii) for every letter on the right-hand side, place one commitment corresponding to that letter. (The two commitments in (ii) will be used to obtain commitments to both an addition result and a carry.)

Let us illustrate how to arrange commitments in Step 6 by considering the puzzle problem shown in Fig. 1 as an example; note that "SEND" is the first term on the left-hand side, "MORE" is the second term, and "MONEY" is the right-hand side. Because letters M, O, R, and E (which constitute the second

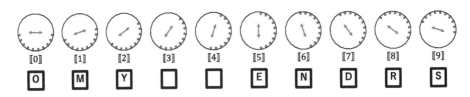

Fig. 14. Example of correspondence between 10 dihedral cards and symbolic cards; P places symbolic cards face-down

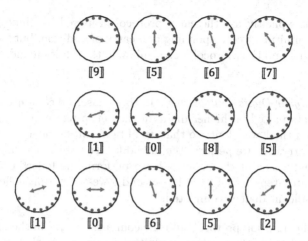

Fig. 15. Example of commitment placement

term) appear twice, twice, once, and thrice, respectively, we apply the copy protocol to obtain three commitments corresponding to "M," three commitments corresponding to "O," two commitments corresponding to "R," and four commitments corresponding to "E." For the other letters, i.e., S, N, D, and Y, we obtain as many commitments as appeared in the problem. After copying, we place the obtained commitments at the corresponding positions on the board, as illustrated in Fig. 15. Note that there should be two commitments corresponding to each of letters M, O, R, and E in the figure. Recall that the blue arrows and dots in Fig. 15 were drawn with invisible ink, and that V does not know any values of the cards.

Addition of Least Significant Digits (Half Adder). In this phase, we compute the addition of the two commitments corresponding to the least significant digits placed in the Setup phase and output commitments to the result of addition and carry (to the higher digits), i.e., we perform the half adder.

1. Utilizing the addition protocol introduced in Sect. 2.3.2, add the value of the commitment corresponding to the least significant digit in the first term (on the left-hand side) to the two commitments corresponding to the least significant digit in the second term. (Recall that two commitments were placed for every letter in the second term.) Therefore, P and V obtain two identical commitments corresponding to the result of the addition.
2. Apply the sign-to-value protocol introduced in Sect. 2.3.4 to one of the two commitments obtained in the previous step to obtain a commitment to the carry to the higher digits.
3. Apply the sign normalization protocol introduced in Sect. 2.3.3 to the other commitment, i.e., the one that was not used in the previous step. (Thanks to this step, any commitment has a value between 0 and $m - 1$.)

We do not explicitly use the carry protocol proposed by Shinagawa [30,31] to obtain a commitment corresponding to a carry of addition, but we do employ the same idea behind the protocol, i.e., combining the addition and sign-to-value protocols.

Addition of Higher Digits (Full Adder). In this phase, we compute the addition of commitments placed on higher digits in the Setup phase and a commitment to a carry from the lower digits so that we obtain commitments to the addition result and a carry, i.e., we perform the full adder. For every pair of higher digits (on the left-hand side of the equation), we perform the following one by one. That is, execute Steps 1 to 4 from the second lowest digit until the addition of the most significant digit is completed.

1. Utilizing the addition protocol, add the commitment $[\![x]\!]$ in the first term (on the right-hand side) to the two commitments $[\![y]\!], [\![y]\!]$ in the second term to obtain two commitments to the result of addition $[\![x+y]\!], [\![x+y]\!]$.
2. Similarly, add the commitment to the carry $[\![c]\!]$ to the two commitments obtained in Step 1 to obtain two commitments to the result of addition with carry $[\![x+y+c]\!], [\![x+y+c]\!]$.
3. Apply the sign normalization protocol to one of the two commitments obtained in Step 2.
4. Apply the sign-to-value protocol to the other commitment to obtain a commitment corresponding to a carry to the higher digit.

Verification. In this phase, we verify whether the rules of Cryptarithmetic are satisfied using the commitments placed in the Setup phase and those obtained in the addition phases.

1. The verifier V checks that every most significant digit is not equal to 0 by partially illuminating the commitment with black light, as shown in Sect. 2.1. This is possible because of the application of the sign normalization protocol in the addition phase. If V finds a value of 0 (for at least one of them), then V rejects it.
2. The verifier V checks that the result of addition is equal to the left-hand side: Apply a rotation shuffle to every pair of commitments that should correspond to the same letter, i.e., a commitment placed in the Setup phase and the one obtained in the addition phase, and then reveal their values. If V finds a pair with different values, then V rejects it.

4   Evaluation

In this section, we demonstrate that our protocol constructed in Sect. 3 works. Specifically, in Sect. 4.1, we count the numbers of cards and shuffles required for our proposed protocol. In Sect. 4.2, we show that our protocol is surely a zero-knowledge proof protocol.

Table 1. Number of cards and shuffles in the proposed protocol

Number of cards	Number of shuffles
$4d + 2$	$7d + 11$

4.1 Numbers of Cards and Shuffles

In this subsection, we evaluate the performance of our proposed protocol, i.e., we count the numbers of required cards and shuffles.

Because the number of symbolic cards is fixed, i.e., it is always 10, we consider only the number of dihedral cards. Regarding the number of shuffles, we consider the worst case. For simplicity, a puzzle instance is assumed to be an equation in which the two terms on the left-hand side have the same number of letters and the right-hand side has one more than that, e.g., the puzzle shown in Fig. 1. We denote the number of letters in the first term on the left-hand side by d. Note that the second term also has d letters and the right-hand side has $d + 1$ letters; hence, the total number of letters in the puzzle instance is $3d + 1$.

Let us count the number of required dihedral cards. After the setup phase, there are $4d+1$ commitments (cards), and to produce such copied commitments, one more card is required during the final copy. Therefore, the protocol uses $4d+2$ cards.

Next, let us count the number of shuffles. For the setup, in addition to the pile-scramble shuffle, the initialization protocol and the copy protocol use a shuffle and they are executed at most 10 times in total. Thus, the total number of shuffles in the setup is 11. To add the least significant digit, the addition, sign-to-value, and sign normalization protocols are performed once each. Thus, the total number of shuffles here is 5. For the addition of the higher digits, the addition is performed twice, sign-to-value and sign normalization protocols are performed once per digit, and the process is repeated for the number of digits to be added. Thus, the total number of shuffles here is $6(d-1)$. For verification, the rotation shuffle is executed for the number of digits of the addition result. Thus, the total number of shuffles in the verification is $d+1$. Therefore, the total number of shuffles in the entire protocol is at most $7d + 11$.

Table 1 shows the number of cards and the number of shuffles derived from the foregoing discussion.

4.2 Proof

In this subsection, we verify that the proposed protocol described in Sect. 3.2 satisfies the zero-knowledge proof property.

Completeness. The verifier V verifies that the values of the 10 dihedral cards are all different, from 0 to 9, in Step 1 of the Setup phase. Because the same commitment is assigned to the same letter and different commitments are assigned to different letters, V is convinced that the same numeral is assigned to the same letter and different numerals are assigned to different letters. Furthermore, if the prover P knows the solution, P places the symbolic card corresponding to each letter of the puzzle instance that satisfies the rules; hence, V is convinced that the solution that P knows satisfies the rule in the Verification phase.

Extractability. As mentioned above, the value of any commitment in the Setup phase is between 0 and 9. If P gives a false input, i.e., a numeral that does not match the solution for a certain letter, V will not be convinced because the solution is assumed to be unique in this study and will output an addition result that is different from the solution, meaning that the value disclosed in Step 2 of the Verification phase will be different.

Zero-Knowledge. Because the commitments corresponding to the letters are prepared using symbolic cards so that V does not know the correspondence, V cannot see which numeral corresponds to which letter in the Setup phase. In the Addition phases, the values of both input and output are not disclosed, so that no information about the solution is leaked to V. In the Verification phase, the values of the commitments are disclosed in Step 2, but V cannot know the information about the original value due to the shuffle operation.

5   Conclusion

In this paper, we proposed a card-based ZKP protocol for Cryptarithmetic using dihedral cards. Our protocol was obtained by constructing copy, half-adder, and full-adder protocols working on dihedral cards with the help of existing basic sub-protocols.

Our future work includes improving the efficiency of the protocol. For example, as shown in Fig. 1, when the number of letters on the right-hand side is larger than that of each of the two terms on the left-hand side, the numeral corresponding to the most significant digit on the right-hand side is automatically determined to be 1. Therefore, the number of shuffles can be reduced by disclosing the commitment to confirm whether it is 1, instead of a rotation shuffle. It would also be interesting to measure the execution time of our protocol in more detail, e.g., [16], or to adopt the "private permutation" model, which allows players' private actions, e.g., [12,18,20,36].

In actual examples [33–35], there are many Cryptarithmetic problems involving the addition of two inputs as well as the addition of three or more inputs. Therefore, one of the tasks is to consider the application to the addition of three or more inputs.

Acknowledgments. We thank the anonymous referees, whose comments have helped us improve the presentation of the paper. We would like to thank Hideaki Sone for his cooperation in preparing a Japanese draft version at an earlier stage of this work. This work was supported in part by JSPS KAKENHI Grant Numbers JP19J21153 and JP21K11881.

References

1. Bultel, X., Dreier, J., Dumas, J.G., Lafourcade, P.: Physical zero-knowledge proofs for Akari, Takuzu, Kakuro and KenKen. In: Demaine, E.D., Grandoni, F. (eds.) Fun with Algorithms. Leibniz International Proceedings in Informatics (LIPIcs), vol. 49, pp. 8:1–8:20. Schloss Dagstuhl, Dagstuhl (2016). https://doi.org/10.4230/LIPIcs.FUN.2016.8

2. Bultel, X., et al.: Physical zero-knowledge proof for Makaro. In: Izumi, T., Kuznetsov, P. (eds.) SSS 2018. LNCS, vol. 11201, pp. 111–125. Springer, Cham (2018). https://doi.org/10.1007/978-3-030-03232-6_8

3. Chien, Y.-F., Hon, W.-K.: Cryptographic and physical zero-knowledge proof: From sudoku to Nonogram. In: Boldi, P., Gargano, L. (eds.) FUN 2010. LNCS, vol. 6099, pp. 102–112. Springer, Heidelberg (2010). https://doi.org/10.1007/978-3-642-13122-6_12

4. Dumas, J.-G., Lafourcade, P., Miyahara, D., Mizuki, T., Sasaki, T., Sone, H.: Interactive physical zero-knowledge proof for Norinori. In: Du, D.-Z., Duan, Z., Tian, C. (eds.) COCOON 2019. LNCS, vol. 11653, pp. 166–177. Springer, Cham (2019). https://doi.org/10.1007/978-3-030-26176-4_14

5. Epstein, D.: On the NP-completeness of cryptarithms. ACM SIGACT News 18(3), 38–40 (1987). https://doi.org/10.1145/24658.24662

6. Goldwasser, S., Micali, S., Rackoff, C.: The knowledge complexity of interactive proof systems. SIAM J. Comput. 18(1), 186–208 (1989). https://doi.org/10.1137/0218012

7. Gradwohl, R., Naor, M., Pinkas, B., Rothblum, G.N.: Cryptographic and physical zero-knowledge proof systems for solutions of Sudoku puzzles. In: Crescenzi, P., Prencipe, G., Pucci, G. (eds.) FUN 2007. LNCS, vol. 4475, pp. 166–182. Springer, Heidelberg (2007). https://doi.org/10.1007/978-3-540-72914-3_16

8. Gradwohl, R., Naor, M., Pinkas, B., Rothblum, G.N.: Cryptographic and physical zero-knowledge proof systems for solutions of Sudoku puzzles. Theor. Comput. Syst. 44(2), 245–268 (2009). https://doi.org/10.1007/s00224-008-9119-9

9. Ishikawa, R., Chida, E., Mizuki, T.: Efficient card-based protocols for generating a hidden random permutation without fixed points. In: Calude, C.S., Dinneen, M.J. (eds.) UCNC 2015. LNCS, vol. 9252, pp. 215–226. Springer, Cham (2015). https://doi.org/10.1007/978-3-319-21819-9_16

10. Lafourcade, P., Miyahara, D., Mizuki, T., Robert, L., Sasaki, T., Sone, H.: How to construct physical zero-knowledge proofs for puzzles with a "single loop" condition. Theor. Comput. Sci. (2021, in press). https://doi.org/10.1016/j.tcs.2021.07.019

11. Lafourcade, P., Miyahara, D., Mizuki, T., Sasaki, T., Sone, H.: A physical ZKP for Slitherlink: How to perform physical topology-preserving computation. In: Heng, S.-H., Lopez, J. (eds.) ISPEC 2019. LNCS, vol. 11879, pp. 135–151. Springer, Cham (2019). https://doi.org/10.1007/978-3-030-34339-2_8

12. Manabe, Y., Ono, H.: Card-based cryptographic protocols for three-input functions using private operations. In: Flocchini, P., Moura, L. (eds.) IWOCA 2021. LNCS, vol. 12757, pp. 469–484. Springer, Cham (2021). https://doi.org/10.1007/978-3-030-79987-8_33

13. Minhaz, A., Singh, A.V.: Solution of a classical cryptarithmetic problem by using parallel genetic algorithm. In: Reliability, Infocom Technologies and Optimization, pp. 1–5. IEEE (2014). https://doi.org/10.1109/ICRITO.2014.7014715

14. Miyahara, D., et al.: Card-based ZKP protocols for Takuzu and Juosan. In: Farach-Colton, M., Prencipe, G., Uehara, R. (eds.) Fun with Algorithms. Leibniz International Proceedings in Informatics (LIPIcs), vol. 157, pp. 20:1–20:21. Schloss Dagstuhl, Dagstuhl (2020). https://doi.org/10.4230/LIPIcs.FUN.2021.20

15. Miyahara, D., Sasaki, T., Mizuki, T., Sone, H.: Card-based physical zero-knowledge proof for Kakuro. IEICE Trans. Fundam. Electron. Commun. Comput. Sci. **102**(9), 1072–1078 (2019). https://doi.org/10.1587/transfun.E102.A.1072

16. Miyahara, D., Ueda, I., Hayashi, Y., Mizuki, T., Sone, H.: Analyzing execution time of card-based protocols. In: Stepney, S., Verlan, S. (eds.) UCNC 2018. LNCS, vol. 10867, pp. 145–158. Springer, Cham (2018). https://doi.org/10.1007/978-3-319-92435-9_11

17. Mizuki, T., Asiedu, I.K., Sone, H.: Voting with a logarithmic number of cards. In: Mauri, G., Dennunzio, A., Manzoni, L., Porreca, A.E. (eds.) UCNC 2013. LNCS, vol. 7956, pp. 162–173. Springer, Heidelberg (2013). https://doi.org/10.1007/978-3-642-39074-6_16

18. Nakai, T., Misawa, Y., Tokushige, Y., Iwamoto, M., Ohta, K.: How to solve millionaires' problem with two kinds of cards. New Gener. Comput. **39**(1), 73–96 (2021). https://doi.org/10.1007/s00354-020-00118-8

19. Nozaki, Y., Hendrian, D., Yoshinaka, R., Shinohara, A.: Enumeration of cryptarithms using deterministic finite automata. In: Câmpeanu, C. (ed.) CIAA 2018. LNCS, vol. 10977, pp. 286–298. Springer, Cham (2018). https://doi.org/10.1007/978-3-319-94812-6_24

20. Ono, H., Manabe, Y.: Card-based cryptographic logical computations using private operations. New Gener. Comput. **39**(1), 19–40 (2021). https://doi.org/10.1007/s00354-020-00113-z

21. Robert, L., Miyahara, D., Lafourcade, P., Mizuki, T.: Interactive physical ZKP for connectivity: Applications to Nurikabe and Hitori. In: De Mol, L., Weiermann, A., Manea, F., Fernández-Duque, D. (eds.) Connecting with Computability. LNCS, vol. 12813, pp. 373–384. Springer, Cham (2021). https://doi.org/10.1007/978-3-030-80049-9_37

22. Ruangwises, S.: An improved physical ZKP for Nonogram (2021). https://arxiv.org/abs/2106.14020

23. Ruangwises, S.: Two standard decks of playing cards are sufficient for a ZKP for Sudoku (2021). https://arxiv.org/abs/2106.13646

24. Ruangwises, S., Itoh, T.: Physical zero-knowledge proof for Numberlink. In: Farach-Colton, M., Prencipe, G., Uehara, R. (eds.) Fun with Algorithms. Leibniz International Proceedings in Informatics (LIPIcs), vol. 157, pp. 22:1–22:11. Schloss Dagstuhl, Dagstuhl (2020). https://doi.org/10.4230/LIPIcs.FUN.2021.22

25. Ruangwises, S., Itoh, T.: Physical zero-knowledge proof for Numberlink puzzle and k vertex-disjoint paths problem. New Gener. Comput. **39**(1), 3–17 (2021). https://doi.org/10.1007/s00354-020-00114-y

26. Ruangwises, S., Itoh, T.: Physical zero-knowledge proof for Ripple Effect. In: Uehara, R., Hong, S.-H., Nandy, S.C. (eds.) WALCOM 2021. LNCS, vol. 12635, pp. 296–307. Springer, Cham (2021). https://doi.org/10.1007/978-3-030-68211-8_24

27. Ruangwises, S., Itoh, T.: Physical ZKP for connected spanning subgraph: applications to Bridges Puzzle and other problems. In: Unconventional Computation and Natural Computation. LNCS, Springer, Cham (2021, to appear)

28. Sasaki, T., Miyahara, D., Mizuki, T., Sone, H.: Efficient card-based zero-knowledge proof for Sudoku. Theor. Comput. Sci. **839**, 135–142 (2020). https://doi.org/10.1016/j.tcs.2020.05.036

29. Sasaki, T., Mizuki, T., Sone, H.: Card-based zero-knowledge proof for Sudoku. In: Ito, H., Leonardi, S., Pagli, L., Prencipe, G. (eds.) Fun with Algorithms. Leibniz International Proceedings in Informatics (LIPIcs), vol. 100, pp. 29:1–29:10. Schloss Dagstuhl, Dagstuhl (2018). https://doi.org/10.4230/LIPIcs.FUN.2018.29

30. Shinagawa, K.: Card-based cryptography with invisible ink. In: Gopal, T.V., Watada, J. (eds.) TAMC 2019. LNCS, vol. 11436, pp. 566–577. Springer, Cham (2019). https://doi.org/10.1007/978-3-030-14812-6_35

31. Shinagawa, K.: Card-based cryptography with dihedral symmetry. New Gener. Comput. 39(1), 41–71 (2021). https://doi.org/10.1007/s00354-020-00117-9

32. Shinagawa, K., et al.: Card-based protocols using regular polygon cards. IEICE Trans. Fundam. Electron. Commun. Comput. Sci. **E100.A**(9), 1900–1909 (2017). https://doi.org/10.1587/transfun.E100.A.1900

33. Tamura, N.: Cryptarithmetic puzzle solver. https://tamura70.gitlab.io/web-puzzle/cryptarithm/. Accessed 21 Mar 2021

34. Torsten, S.: Alphametics and cryptarithms. https://www.math.uni-bielefeld.de/~sillke/PUZZLES/ALPHAMETIC/. Accessed 21 Mar 2021

35. Truman, C.: Alphametic puzzles. http://www.tkcs-collins.com/truman/alphamet/alphamet.shtml, Accessed 21 Mar 2021

36. Yasunaga, K.: Practical card-based protocol for three-input majority. IEICE Trans. Fundam. Electron. Commun. Comput. Sci. **E103.A**(11), 1296–1298 (2020). https://doi.org/10.1587/transfun.2020EAL2025

Quantum Algorithm for Dyck Language with Multiple Types of Brackets

Kamil Khadiev[1,2]([✉]) and Dmitry Kravchenko[3]

[1] Kazan Federal University, Kazan, Russia
[2] Kazan E. K. Zavoisky Physical-Technical Institute, Kazan, Russia
[3] Center for Quantum Computer Science, Faculty of Computing,
University of Latvia, Riga, Latvia

Abstract. We consider the recognition problem of the Dyck Language generalized for multiple types of brackets. We provide an algorithm with quantum query complexity $O(\sqrt{n}(\log n)^{0.5k})$, where n is the length of input and k is the maximal nesting depth of brackets. Additionally, we show the lower bound for this problem which is $\Omega(\sqrt{n}c^k)$ for some constant c.

Interestingly, classical algorithms solving the Dyck Language for multiple types of brackets substantially differ from the algorithm solving the original Dyck language. At the same time, quantum algorithms for solving both kinds of the Dyck language are of similar nature and requirements.

Keywords: Dyck language · Regular language · Strings · Quantum algorithms · Query complexity

1 Introduction

Quantum computing [2,3,19] is one of the hot topics in computer science of the last decades. There are many problems where quantum algorithms outperform the best known classical ones [13], and one of the most important performance metrics in this regard is *query complexity*. We refer to [3] for a nice survey on the quantum query complexity, and to [10,14–18] for the more recent progress.

Among other problems, quantum technologies can reduce the query complexity of recognizing many formal languages [5]. In this paper we consider a problem of recognizing whether an n-character string belongs to one important *regular language*. Although this problem may seem too specific, we believe our approach to model a variety of computational tasks that can be described by regular languages.

Aaronson, Grier and Schaeffer [1] have recently shown that any regular language L may have one of three possible quantum query complexities on inputs of length n: $\Theta(1)$ if the language can be decided by looking at $O(1)$ first or last symbols of a word; $\tilde{\Theta}(\sqrt{n})$ if the best way to decide L is Grover's search (for example, for the language consisting of all words containing at least one letter

© Springer Nature Switzerland AG 2021
I. Kostitsyna and P. Orponen (Eds.): UCNC 2021, LNCS 12984, pp. 68–83, 2021.
https://doi.org/10.1007/978-3-030-87993-8_5

a); $\Theta(n)$ for languages in which one can embed counting modulo some number p which has quantum query complexity $\Theta(n)$ (for example, the binary XOR function).

s As shown in [1], a regular language being of complexity $\tilde{O}(\sqrt{n})$ (which includes the first two cases of the list above) is equivalent to it being *star-free*. Star-free languages are defined as the languages which have regular expressions not containing the Kleene star (if it is allowed to use the complement operation). Star-free languages are one of the most commonly studied subclasses of regular languages and there are many equivalent characterizations of them.

One class of the star-free languages mentioned in [1] is the Dyck languages (with one type of brackets and with constant height k). To introduce a brief intuition about these languages, we may mention that words "[ ]" and "[[ ] [ ]]" belong to a Dyck language, while words "] [" and "[ ]] [[ ]" do not. Formally, Dyck language with height k consists of all words with balanced number of brackets such that in no prefix the number of opening brackets exceeds the number of closing brackets by more than k; we denote the problem of determining if an input of length n belongs to this language by $\text{DYCK}_{k,n}$. We note that such language of unbounded height (i.e. $k = \frac{n}{2}$) is a fundamental example of a context-free language that is not regular.

For this problem, Ambainis et al. [4] show that an exponential dependence of the complexity on k is unavoidable. Namely, for the balanced brackets language (i) there exists $c > 1$ such that, for all $k \leq \log n$, the quantum query complexity is $\Omega(c^k \sqrt{n})$; (ii) if $k = c \log n$ for an appropriate constant c, then the quantum query complexity is $\Omega(n^{1-\epsilon})$.

Thus, the exponential dependence on k is unavoidable and distinguishing sequences of balanced brackets of length n and depth $\log n$ is almost as hard as distinguishing sequences of length n and arbitrary depth. Similar lower bounds have recently been independently proven by Buhrman et al. [8]. Additionally, Ambainis et al. [4] describe an explicit algorithm for the decision problem $\text{DYCK}_{k,n}$ with $O\left(\sqrt{n}(\log n)^{0.5k}\right)$ quantum queries. The algorithm also works for arbitrary k, and outperforms the trivial upper bound of n when $k = o\left(\frac{\log n}{\log \log n}\right)$.

This work generalizes $\text{DYCK}_{k,n}$ to the case of multiple types of brackets. For example, such languages contain words like "[( )]" and do not contain words like "[( ])" (here square and round brackets are the two different types of brackets). We denote the problem of determining if an input of length n belongs to the Dyck language of height k and at most t types of brackets by $\text{DYCK}_{k,n,t}$. Obviously, $\text{DYCK}_{k,n,1} = \text{DYCK}_{k,n}$.

We note that $\text{DYCK}_{k,n}$ and $\text{DYCK}_{k,n,t}$ for $t > 1$ are two substantially different problems regarding classical (deterministic or randomized) calculations. The former problem allows using a counter to keep the number of currently open brackets and thus be content with the memory size of $O(\log k)$. In contrast, the latter problem requires keeping all the sequence of currently open brackets in a stack, which may take up to $O(k)$ memory. While both problems are solvable in linear time, there is an exponential gap in the memory usage.

In this paper we provide a quantum algorithm for $\text{DYCK}_{k,n,t}$ with quantum query complexity $O(\sqrt{n}(\log n)^{0.5k})$. We apply the known technique of solving $\text{DYCK}_{k,n}$, and then perform a more complex but slightly faster procedure to check the type-matching of the brackets.

The structure of the paper is the following. Section 2 describes some conventional notions for quantum computation. Section 3 provides the main algorithm and the proofs, with the discussion on the complexity of the algorithm and on the lower bounds given in Subsect. 3.3. In the final Sect. 4 we introduce several attempts and wishes to make our techniques more general and our evaluations more precise.

2   Preliminaries

2.1   Definitions

We use the following formalism throughout the paper. We assume an input string to consist of brackets of t *types* for some positive integer t; each type is represented by a pair of brackets – an opening and a closing one. Further, we assume the brackets to be encoded by integers from 1 to $2t$, where the opening and the closing brackets of i-th type correspond to the numbers $2i - 1$ and $2i$ respectively.

We define two functions:

- Function $Type : \{1, \ldots, 2t\} \to \{1, \ldots, t\}$ returns the type of a bracket. $Type(x) = \lceil x/2 \rceil$.
- Function $Open : \{1, \ldots, 2t\} \to \{0, 1\}$ returns 1 if the argument is an opening bracket, or 0 if it is a closing bracket. $Open(x) = x \mod 2$.

For example, string $S = $ "[( )]" could be encoded as "$1, 3, 4, 2$". Then

$$Type(s_1) = Type(s_4) = 1 \quad \text{stand for the square brackets;}$$
$$Type(s_2) = Type(s_3) = 2 \quad \text{– for the parentheses;}$$
$$Open(s_1) = Open(s_2) = 1 \quad \text{– for the opening brackets; and}$$
$$Open(s_3) = Open(s_4) = 0 \quad \text{– for the closing brackets.}$$

We call a string $S = (s_1, \ldots, s_m)$ a *well-balanced* sequence of brackets if one of the following holds:

1. S is empty;
2. S consists of two well-balanced subsequent substrings, i.e. $S[1, i]$ and $S[i + 1, m]$ are both well-balanced for some i (hereafter we denote by $S[i, j]$ a substring $(s_i, \ldots, s_j)$ of a string $S = (s_1, \ldots, s_m)$);
3. S is a correctly embraced well-balanced sequence, i.e.
   - $S[2, m - 1]$ is a well-balanced sequence,
   - $Type(s_1) = Type(s_m)$,
   - $Open(s_1) = 1$ and $Open(s_m) = 0$.

Obviously, the set of all well-balanced sequences of brackets defines the DYCK language.

We also introduce a metric for the balancedness of a substring. Let f be a function which returns the difference between the numbers of opening and closing brackets: $f(S[l,r]) = \#_1(S[l,r]) - \#_0(S[l,r])$. (Here $\#_x(S[l,r])$ denotes the number of symbols s_j, for $l \leq j \leq r$, such that $Open(s_j) = x$). We define a $+k$-substring (resp. $-k$-substring) as a substring whose balance is equal to k (resp. equal to $-k$). A $\pm k$-substring is a substring whose balance is equal to k in absolute value.

We call a nonempty substring $S[l,r]$ *minimal* if it does not contain a nonempty substring $S[l',r']$ such that $(l,r) \neq (l',r')$ and $f(S[l',r']) = f(S[l,r])$. We call a nonempty substring $S[l,r]$ *prefix-minimal* if it does not start with $S[l,r']$ such that $r' < r$ and $f(S[l,r']) = f(S[l,r])$. We define the *height* of a substring $S[l,r]$ as $h(S[l,r]) = \max_{i \in \{l,\dots,r\}} f(S[l,i])$.

For example, string $S =$ "[ ] ( )" is well-balanced, because it consists of two well-balanced substrings "[ ]" and "( )", which in turn both are correctly embraced empty strings. Its substring $S[1,2] =$ "[ ]" is both minimal and prefix-minimal, whereas its substring $S[2,4] =$ "] ( )" is neither minimal nor prefix-minimal (since $f(S[2,2]) = f(S[2,4]) = -1$).

Finally, we define the problem $\text{DYCK}_{k,n,t}(S)$. Function $\text{DYCK}_{k,n,t}$ accepts $S = (s_1, \dots, s_n)$ as an input and

- returns 1 if S is a well-balanced sequence of brackets with at most t types of brackets and with $h(S) \leq k$;
- returns 0 otherwise.

2.2 Computational Model

To evaluate the complexity of a quantum algorithm, we use the standard form of the quantum query model [3]. It is a generalization of the decision tree model of classical computation that is commonly used to lower bound the amount of time required for a computation.

Let $g : D \to \{0,1\}$, for some $D \subseteq \{0,1\}^n$, be an n-argument binary function we wish to compute. We have an oracle access to the input x—it is implemented by a specific unitary transformation usually defined as $|i\rangle|z\rangle|w\rangle \to |i\rangle|z \oplus x_i\rangle|w\rangle$, where the $|i\rangle$ register indicates the index of the variable we are querying, $|z\rangle$ is the output register, and $|w\rangle$ is some auxiliary work-space. An algorithm in the query model consists of alternating applications of arbitrary unitaries independent of the input and the query unitary, and a measurement in the end. The smallest number of queries for an algorithm that outputs $g(x)$ with probability $\geq \frac{2}{3}$ on all x is called the quantum query complexity of the function g and is denoted by $Q(g)$. If the error probability stays $\leq \frac{1}{3}$ for any input and output, the algorithm is said to have *two-side bounded error* probability (as opposed to TRUE- or FALSE-biased algorithms which are said to have one-side bounded error).

Throughout this paper, by the running time of an algorithm we mean a number of queries to an oracle. In particular, we assume the oracle to process queries *Type* and *Open* in constant time.

More information on quantum computation and query model can be found in [2,3,19].

To distinguish ordinary deterministic and randomized complexities from the quantum complexity, they are traditionally called by one term *classical complexity.*

3   Quantum Algorithm

Before introducing the algorithm for solving $\text{DYCK}_{k,n,t}(S)$, we mention the following result from [4], which will be used as important subroutine.

Lemma 1 ([4], **Theorem 3**). *There exists a quantum algorithm that solves* $\text{DYCK}_{k,n,1}$ *in time* $O(\sqrt{n}(\log n)^{0.5k})$. *The algorithm has two-side bounded error probability* $\varepsilon < 0.5$.

The algorithm for solving $\text{DYCK}_{k,n,t}(S)$ generally consists of three main steps:

Step 1. Check whether there are at most t types of brackets, and return 0 if the number of types exceeds t. This part is discussed in Sect. 3.1.

Step 2. Uniformize S to just one type of brackets by considering a string $Y = (y_1, \ldots, y_n)$ where $y_i = Open(s_i)$. Check whether $\text{DYCK}_{k,n,1}(Y) = 1$ by using the algorithm from Lemma 1. If this is the case, then S is a well-balanced sequence of brackets with their types ignored. Otherwise, S obviously is not well-balanced and $\text{DYCK}_{k,n,t}(S) = 0$. This step almost exactly repeats the algorithm from [4].

Step 3. Check whether for any substring $S[l,r]$ the following condition holds: If $Y[l,r]$ is a well-balanced sequence of brackets (with their types ignored) of depth v and $Y[l+1,r-1]$ is a well-balanced sequence of brackets of depth $v-1$, then (i) $Type(s_l) = Type(s_r)$; and (ii) $S[l+1,r-1]$ is a well-balanced sequence of brackets.

Step 3 should be considered as the main contribution of the paper, and we describe it in detail in Sect. 3.2. By the definition of the problem, if S passes all three checks, then $\text{DYCK}_{k,n,t}(S) = 1$. The complexity of the problem is evaluated in Sect. 3.3, and in Sect. 3.4 we summarize our approach to one formal listing.

3.1   Procedure for Step 1

Recall that by the assumption, all the brackets are encoded by integers from 1 to $2t$. Hence it only remains to check whether S contains a bracket with code $c > 2t$. This problem obviously can be solved by Grover's algorithm [6,11] for finding an argument j (if any) such that $g(j) = 1$, for an arbitrary function $g : \{1, \ldots, n\} \to \{0, 1\}$ implemented as a quantum oracle. To apply this technique for solving Step 1, it is sufficient to consider the binary function $g(j)$ which returns 1 iff $s_j > 2t$. Grover's algorithm runs in time $O(\sqrt{n})$ and has error probability at most 0.5; and so are the complexity and the error probability of Step 1.

The assumption on encoding of brackets could be relaxed by allowing to use any distinct integer for each kind of bracket. Then the problem becomes more complex: determine whether a sequence $s_1, \ldots, s_n$ contains at most $2t$ distinct integers. The upper bound for its query complexity is $O(\sqrt{nt} \log t)$. We refer to Sect. 4.1 for more details.

3.2 Procedure for Step 3

Assume that any 0-substring $S[l', r']$ with $h(S[l', r']) \leq v - 1$ is known to be a well-balanced sequence of brackets. In this section we present a procedure that checks whether, under this assumption, any 0-substring $S[l, r]$ with $h(S[l, r]) = v$ is a well-balanced sequence of brackets.

We wish to implement a function CHECKSUBSTR(S, v) which returns

- TRUE if there exists a "wrong" (not well-balanced) sequence $S[l, r]$ such that $h(S[l, r]) = v$;
- FALSE otherwise.

If we had CHECKSUBSTR(S, v) implemented, then we could invoke it for each $v \in \{1, \ldots, k\}$. In case of all-FALSE output, the function should return FALSE ("no wrong sequences"), otherwise TRUE ("found a wrong sequence for at least one height $v \in \{1, \ldots, k\}$").

We propose the following implementation of CHECKSUBSTR(S, v).

The Case $v = 1$. We start with considering the case $v = 1$. Let a function $g^1 : \{1, \ldots, n-1\} \to \{0, 1\}$ be such that $g^1(j) = 1$ iff $Open(s_j) = 1$, $Open(s_{j+1}) = 0$, and $Type(s_j) \neq Type(s_{j+1})$. In other words, the function indicates sequentially opening and closing brackets of different types.

We use Grover's algorithm to search for an argument $j \in \{1, \ldots, n\}$ such that $g^1(j) = 1$. Hereafter we call this subroutine GROVER$(g^1, 1, n)$, where g^1 is the function run by a quantum oracle in constant time, and $1 \ldots n$ defines an interval to search in. If GROVER$(g^1, 1, n)$ finds such index j, then CHECKSUBSTR$(S, 1)$ returns TRUE, otherwise FALSE.

Note that due to the complexity of Grover's algorithm, the query complexity of GROVER$(g^1, 1, n)$ is $O(\sqrt{n})$, with the error probability at most 0.5.

The Case $v > 1$. This step allows assuming any 0-substring $S[l', r']$ with $h(S[l', r']) = v-1$ to be a well-balanced sequence of brackets. Under this assumption, we show that the next property holds:

Lemma 2. *If for an input string S, any 0-substring $S[l', r']$ with $h(S[l', r']) = v-1$ is a well-balanced sequence of brackets, then any prefix-minimal 0-substring $S[l, r]$ with $h(S[l, r]) = v$ is such that $S[l+1, r-1]$ is either empty or a well-balanced sequence of brackets.*

Proof. According to the definition of a prefix-minimal 0-substring, we claim that $S[l, r]$ does not contain any shorter prefix 0-substring. In particular, it means that $Open(s_l) = 1$ and $Open(s_r) = 0$. Therefore, $h(S[l+1, r-1]) = v - 1$, and $S[l+1, r-1]$ is a 0-substring. Due to the assumption of the lemma, $S[l+1, r-1]$ is a well-balanced sequence of brackets. □

Therefore, to complete checking whether the 0-substring $S[l, r]$ with $h(S[l, r]) = v$ is a well-balanced sequence of brackets, it only remains to check that $Type(s_l) = Type(s_r)$.

Let us present a subroutine that searches for a 0-substring $S[l, r]$ with $h(S[l, r]) = v$ such that $Type(s_l) \neq Type(s_r)$. If this subroutine finds nothing, it means that any 0-substring $S[l, r]$ with $h(S[l, r]) = v$ is well-balanced.

We use the following property of prefix-minimal 0-substrings:

Lemma 3. *For any prefix-minimal 0-substring $S[l, r]$ with $h(S[l, r]) = v$, there exist indices r' and l' such that*

- $l \leq r' < l' \leq r$,
- $S[l, r']$ *is a +v-substring,*
- $S[l', r]$ *is a −v-substring, and*
- *there are no ±v-substrings contained in $S[r' + 1, l' - 1]$.*

Proof. Assume that there is no such index $r' \in \{l, \ldots, r - 1\}$ that $S[l, r']$ is a +v-substring. Then we consider the index $j = \text{argmax}_{j \in \{l+1, \ldots, r\}} f(S[l, j])$ and note that $h(S[l, r]) = v$ implies $f(S[l, j]) = v$, which contradicts the assumption. We conclude the that the desired index r' exists.

Now assume that there is no such index $l' \in \{r' + 1, \ldots, r\}$ that $S[l', r]$ is a −v-substring. Recall that by the definition of a 0-substring, $f(S[l, r]) = 0$. At the same time, $f(S[l, r]) = f(S[l, r']) + f(S[r' + 1, r])$ and $f(S[l, r']) = v$. Therefore, $f(S[r' + 1, r]) = f(S[l, r]) - f(S[l, r']) = 0 - v = -v$, which contradicts the assumption. We conclude that both desired indices r' and l' exist.

Finally, assume sequence $S[r' + 1, l' - 1]$ to contain a ±v-substring. Then we consider the leftmost ±v-substring $S[l'', r'']$, where $r' < l'' \leq r'' < l'$.

If $S[l'', r'']$ is a +v-substring, i.e. $f(S[l'', r'']) = v$, then the minimality of l'' implies $f(S[r' + 1, l'' - 1]) > -v$. Then,

$$f(S[l, r'']) = f(S[l, r']) + f(S[r' + 1, l'' - 1]) + f(S[l'', r''])$$
$$= v + f(S[r' + 1, l'' - 1]) + f(S[l' + 1, l'' - 1]) > v$$

contradicts the fact that $h(S[l, r]) = \max_{j \in \{l+1, r\}} f(S[l, j]) = v$.

To finish the proof, it remains only to consider (the impossibility of) the case where $S[l'', r'']$ is a −v-substring, i.e. $f(S[l'', r'']) = -v$. In this case $f(S[r' + 1, l'' - 1])$ can be negative, zero, or positive.

- If $f(S[r' + 1, l'' - 1]) < 0$, then

$$f(S[l, r'']) = f(S[l, r']) + f(S[r' + 1, l'' - 1]) + f(S[l'', r''])$$
$$= v + f(S[r' + 1, l'' - 1]) - v = f(S[r' + 1, l'' - 1]) < 0.$$

Therefore, there exists such index j that $j < r'' < r$ and $f(S[l,j]) = 0$, which contradicts the prefix-minimality of the 0-substring $S[l,r]$.

- If $f(S[r'+1,l''-1]) = 0$, then
$f(S[l,r'']) = f(S[l,r']) + f(S[r'+1,l''-1]) + f(S[l'',r'']) = 0$.
Therefore, $f(S[l,r'']) = 0$ where $r'' < r$, which contradicts the prefix-minimality of the 0-substring $S[l,r]$.

- If $f(S[r'+1,l''-1]) > 0$, then
$f(S[l,l''-1]) = f(S[l,r']) + f(S[r'+1,l''-1]) = v + f(S[r'+1,l''-1]) > v$
contradicts the fact that $h(S[l,r]) = \max_{j \in \{l+1,r\}} f(S[l,j]) = v$.

$\square$

These lemmas allow to formulate the algorithm for searching for a not well-balanced 0-substring, with its length limited to be at most d:

Step 1. Pick index b uniformly at random in $\{1, \dots, n\}$.

Step 2. Search for the leftmost $\pm v$-substring with length at most d, in $S[b, \min(n, b+d-1)]$. If such substring $S[i_r, j_r]$ was found, continue to Step 3. Otherwise proceed to Step 4.

Step 3. Search for the rightmost $\pm v$-substring with length at most d in $S[\max(i_r - d, 1), i_r - 1]$. If such substring $S[i_l, j_l]$ was found, proceed to Step 6. Otherwise stop and return FALSE.

Step 4. Search for the rightmost $\pm v$-substring with length at most d in $S[\max(b - d+1, 1), b]$. If such substring $S[i_l, j_l]$ was found, proceed to Step 5. Otherwise stop and return FALSE.

Step 5. Search for the leftmost $\pm v$-substring with length at most d in $S[j_l + 1, \min(n, j_l + d)]$. If such substring $S[i_r, j_r]$ was found, continue to Step 6. Otherwise stop and return FALSE.

Step 6. If $f(S[i_l, j_l]) > 0$, $f(S[i_r, j_r]) < 0$ and $Type(i_l) \neq Type(j_r)$, then return the resulting substring $S[i_l, j_r]$. Otherwise return FALSE.

To search for the rightmost $\pm v$-substring or for the leftmost $\pm v$-substring of length at most d in a segment, we use a subroutine from [4] with the following property:

Lemma 4 ([4], **Property 2**). *There is a quantum algorithm for searching for the leftmost or for the rightmost $\pm v$-substring of length at most d, in a substring $S[l,r]$. The query complexity of the algorithm is $O(\sqrt{r-l}(\log(r-l))^{0.5(v-2)})$. It returns (i, j, σ) such that $S[i,j]$ is a $\pm v$-substring and $\text{sign}(f(S[i,j])) = \sigma$. It returns FALSE if such substring does not exist.*

Hereafter we call subroutines for the leftmost and for the rightmost $\pm v$-substring respectively LEFTMOST(S, l, r, v, d) and RIGHTMOST(S, l, r, v, d). They return a triple (i, j, σ), such that $S[i,j]$ is the resulting substring and $\sigma = sign(f(S[i,j]))$. They return FALSE if there are no such $\pm v$-strings.

We formalize the algorithm in the code listing of Algorithm 1 that implements a subroutine which we have called CHECKSUBSTRFIXEDLENGTH(S, v, d):

Algorithm 1. Implementation of CHECKSUBSTRFIXEDLENGTH(S, v, d) subroutine, search for a not well-balanced 0-substring $S[l, r]$ with height $h(S[l, r]) = v$ and length $r - l + 1 \leq d$.

$\{1, \ldots, n\} \xleftarrow{R} b$ ▷ randomly pick b

$u_r = (i_r, j_r, \sigma_r) \leftarrow$ LEFTMOST$(S, b, \min(n, b + d - 1), v, d)$

if $u_l \neq$ FALSE **then**

    $u_l = (i_l, j_l, \sigma_l) \leftarrow$ RIGHTMOST$(S, \max(i_r - d, 1), i_r - 1, v, d)$

else

    $u_l = (i_l, j_l, \sigma_l) \leftarrow$ RIGHTMOST$(S, \max(b - d + 1, 1), b, v, d)$

    if $u_l \neq$ FALSE **then**

        $u_r = (i_r, j_r, \sigma_r) \leftarrow$ LEFTMOST$(S, j_l + 1, (n, j_l + d), v, d)$

    end if

end if

if $u_l \neq$ FALSE and $u_r \neq$ FALSE and $\sigma_l = 1$ and $\sigma_r = -1$ and $Type(s_{i_l}) \neq Type(s_{j_r})$

then

    return (i_l, j_r)

else

    return FALSE

end if

Assume that some string S contains a not well-balanced 0-substring $S[l, r]$ with height $h(S[l, r]) = v$ and length d. The probability of finding such substring by this algorithm is equal to the probability of picking an index inside the substring, and therefore can be estimated by $\Omega(d/n)$. By applying the Amplitude amplification algorithm [7] for the randomized Algorithm 1, we obtain an algorithm with query complexity $O(\sqrt{\frac{n}{d}} \cdot \sqrt{d}(\log d)^{0.5(v-2)}) = O(\sqrt{n}(\log d)^{0.5(v-2)})$.

Next, we search for d among the elements of set $T = \{2^0, 2^1, 2^2, \ldots, 2^{\lceil \log_2 n \rceil}\}$. This can be done also by using Grover's algorithm. The overall complexity of the algorithm for finding a 0-substring $S[l, r]$ with height $h(S[l, r]) = v$ and arbitrary length is $O(\sqrt{n}(\log n)^{0.5(v-1)})$. We note that Grover's algorithm relies on an oracle with a two-side bounded error, whereas it is hardly justified to assume a quantum oracle which directly handles T to markup the appropriate lengths. To address this issue, we use the modification of the algorithm presented in [4,12] and thus obtain the implementation of CHECKSUBSTR(S, v). We formulate it in the listing of Algorithm 2 that uses Grover's algorithm implementation from [4,12] as a subroutine GROVERWITHPROBABILISTICORACLE(g, D), where $g : D \rightarrow \{0, 1\}$ is a function with its domain D being the search space. We assume this subroutine to return the target argument if it finds any or FALSE if it fails (just like the standard implementation GROVER(g, D) [6,11] used previously in Step 1). We denote by $g_{S,v} : \{0, \ldots, \lceil \log_2 n \rceil\} \rightarrow \{0, 1\}$ a function such that $g_{S,v}(u) = 1$ iff CHECKSUBSTRFIXEDLENGTH$(s, v, 2^u) \neq$ FALSE.

Algorithm 2. Implementation of CHECKSUBSTR(S, v) subroutine, search for a not well-balanced 0-substring $S[l, r]$ with height $h(S[l, r]) = v$ and any length. $g_{S,v}(u) = 1 \iff$ CHECKSUBSTRFIXEDLENGTH$(s, v, 2^u) \neq$ FALSE

$u \leftarrow$ GROVERWITHPROBABILISTICORACLE$(g_{S,v}, \{0, \ldots, \lceil \log_2 n \rceil\})$
if $u =$ FALSE **then**
    return FALSE
else
    return CHECKSUBSTRFIXEDLENGTH$(s, v, 2^u)$ ▷ either (l, r) or FALSE
end if

Finally, we implement Step 3 in the code listing of Algorithm 3.

Algorithm 3. STEP3(S)

$v \leftarrow 1$
while $v \leq k$ **do**
    if CHECKSUBSTR$(S, v) \neq$ FALSE **then**
        return TRUE
    end if
    $v \leftarrow v + 1$
end while
return FALSE

Then the overall algorithm for the problem DYCK$_{n,k,t}$ combines the three steps in the code listing of Algorithm 4.

Algorithm 4. Solving DYCK$_{n,k,t}$

---

if STEP1$(S) = 1$ **and** DYCK$_{n,k}(Y) = 1$ **and** STEP3$(S) =$ FALSE **then**
    return 1
else
    return 0
end if

3.3 Query Complexity

In this section we estimate the query complexity of DYCK$_{k,n,t}$ and discuss properties of Algorithm 4.

Theorem 1. *Algorithm 4 for solving* DYCK$_{k,n,t}$, *has query complexity* $O(\sqrt{n}(\log n)^{0.5k})$ *and a constant two-side bounded error probability* $\varepsilon < 0.5$.

Proof. We start with the query complexity of the algorithm.

The complexity of Step 1 is obviously equal to the one of Grover's algorithm, i.e. to $O(\sqrt{n})$. Lemma 1 estimates the complexity of Step 2 as $O(\sqrt{n}(\log n)^{0.5k})$. The complexity of Step 3 can be derived from the code listing of Algorithm 3:
$$O(\sum_{v=1}^{k} \sqrt{n}(\log n)^{0.5(v-1)}) = O(\sqrt{n}(\log n)^{0.5(k-1)}).$$

The overall complexity of Algorithm 4 is
$$O(\sqrt{n}) + O(\sqrt{n}(\log n)^{0.5k}) + O(\sqrt{n}(\log n)^{0.5(k-1)}) = O(\sqrt{n}(\log n)^{0.5k}).$$

We continue the proof by considering the error probability of the algorithm. Step 1 has error probability at most 0.5. Step 2 has constant error probability $\varepsilon_0 < 0.5$. Step 3 has error probability at most $1-(1-\varepsilon_1)^k$ for some constant $\varepsilon_1 < 0.5$. As each error probability is constant, we can obtain the desired overall error probability ε by exploiting the technique from [4], i.e. by a series of repetitive calls of the algorithm.   □

We finish our discussion with a couple of lower bounds of the query complexity.

Theorem 2. *There exists a constant $c_1 > 0$ such that* $Q(\text{DYCK}_{c_1 \ell,n,t}) = \Omega(2^{\frac{\ell}{2}}\sqrt{n})$.

Proof. The similar bound holds for $Q(\text{DYCK}_{c_1 \ell,n,1})$ [4, Theorem 6]. Ignoring the types of the brackets makes $\text{DYCK}_{c_1 \ell,n,t}$ equivalent to $\text{DYCK}_{c_1 \ell,n,1}$, therefore $\text{DYCK}_{c_1 \ell,n,t}$ is at least as hard as $\text{DYCK}_{c_1 \ell,n,1}$.   □

Theorem 3. *For any $\gamma > 0$, there exists a constant $c_2 > 0$ such that*
$$Q(\text{DYCK}_{c_2 \log n,n,t}) = \Omega(n^{1-\gamma}).$$

Proof. The similar bound holds for $Q(\text{DYCK}_{c_2 \log n,n,1})$ that was presented in [4, Theorem 5]. Ignoring the types of the brackets makes $\text{DYCK}_{c_2 \log n,n,t}$ equivalent to $\text{DYCK}_{c_2 \log n,n,1}$, therefore $\text{DYCK}_{c_2 \log n,n,t}$ is at least as hard as $\text{DYCK}_{c_2 \log n,n,1}$.   □

3.4   The Overall Algorithm

In this section we aim to provide a general picture of the proposed algorithm. With this purpose, we compile all the fragments from above to one listing in Algorithm 5. Although this could seem redundant, we believe it to be useful for better comprehension, regardless of whether a reader was able to follow the content of this paper up to this point.

Algorithm 5. Overall algorithm for solving $\text{DYCK}_{n,k,t}$

Input: A string $S \in \mathbb{N}^n$.
Output: 1 if $\text{DYCK}_{n,k,t}(S) = 1$ and 0 otherwise

$\qquad\qquad\qquad\qquad\qquad\qquad\qquad\qquad\quad$ ▷ Step 1: $O(\sqrt{n})$ [11]
if $\text{GROVER}(g, \{1, \ldots, n\}) \neq \text{FALSE}$ **then** $\qquad$ ▷ Search for a bracket code $s_i > 2t$
$\qquad$ **return** 0
end if

$\qquad\qquad\qquad\qquad\qquad\qquad\qquad\qquad$ ▷ Step 2: $O(\sqrt{n}(\log n)^{0.5k})$ [4]
if $\text{DYCK}_{n,k}(Y) \neq 1$ **then** $\qquad$ ▷ Check if $S \in \text{DYCK}_{n,k}$ with bracket types ignored
$\qquad$ **return** 0
end if

$\qquad\qquad\qquad\qquad\qquad\qquad\qquad\qquad$ ▷ Step 3: $O(\sqrt{n}(\log n)^{0.5(k-1)})$
for $v \leftarrow 1$ to k **do** $\qquad\qquad$ ▷ Run CHECKSUBSTR for each height up to k
$\qquad$ $u \leftarrow \text{GROVERWITHPROBABILISTICORACLE}(g_{S,v}, \{0, \ldots, \lceil \log_2 n \rceil\})$ $\quad$ ▷ See [4, 12]
$\qquad$ **if** $u \neq \text{FALSE}$ **then**
$\qquad\qquad$ **if** $\text{CHECKSUBSTRFIXEDLENGTH}(s, v, 2^u) \neq \text{FALSE}$ **then** $\qquad$ ▷ See Algorithm 1
$\qquad\qquad\qquad$ **return** 1
$\qquad\qquad$ **end if**
$\qquad$ **end if**
end for

return 0

---

4 Future Work

4.1 Generalizing Step 1

The restriction for all the brackets to be encoded by positive integers up to $2t$, is quite significant for the proposed quantum algorithm. In contrast, formulation of a more natural problem could assume arbitrary encoding of different kinds of brackets. For example, a string could consist of brackets like "( )", "[ ]", "{ }" in arbitrary encoding like ASCII, UTF-32, etc. Under these circumstances, one still can distinguish the type of a certain bracket; and still can determine whether a certain bracket is opening or closing; but one cannot anymore determine how many different types of brackets occur in the string.

Formally speaking, the fragment "at most t types of brackets" from our definition of $\text{DYCK}_{k,n,t}$ means $|\{Type(s_i) : 1 \leq i \leq n\}| \leq t$ rather than $\max_{i \in \{1,\ldots,n\}} Type(s_i) \leq t$ which was assumed throughout the paper. Hereafter we refer to such a more general formulation of the problem as $\text{DYCK}'_{k,n,t}$. The implementation of Step 1 from Sect. 3.1 is not suitable for solving $\text{DYCK}'_{k,n,t}$, whereas the rest of the algorithm does not depend on whether the codes of the types of brackets are consecutive or not.

We note that in many cases this won't be an issue, as the number of different types of brackets t typically is a small constant like 2, 3 or 4. However the following problem could be of certain interest even if not connected with $\text{DYCK}'_{k,n,t}$:

Problem 1. Given a string S of length n, and a limitation parameter t, determine whether $|\{Type(s_i) : 1 \leq i \leq n\}| \leq t$.

Note that Step 1 from Sect. 3.1 obviously reduces to Problem 1.

In the rest of this section we propose an algorithm for solving this problem and thus generalize our solution to $\text{DYCK}'_{k,n,t}$, i.e. to the case with arbitrarily encoded sequences of brackets.

Let T be an integer such that $2T$ is an upper bound for the code of a bracket in the input string (e.g. the size of the input alphabet). Let $Type : \{1, \ldots, 2T\} \rightarrow \{1, \ldots, t\}$ be a function that returns the type of a bracket. Let $q : \{1, \ldots, n\} \times \{1, \ldots, 2T+1\} \rightarrow \{0, \ldots, 2T\}$ be a function which returns

- $q(i,r) = Type(i)$ if $Type(i) < r$; or
- $q(i,r) = 0$ otherwise.

We consider the following procedure:

Step 1 Compute $y_1 = max\{q(i, 2T+1), 1 \leq i \leq n\}$ by using Dürr's and Høyer's algorithm for finding the maximum [9]. Thus we compute the maximum among all the codes of brackets.
Step 2 Compute $y_2 = max\{q(i, y_1), 1 \leq i \leq n\}$ in the same manner, the second-biggest code among all the codes of brackets.
......
Step j Compute $y_j = max\{q(i, y_{j-1}), 1 \leq i \leq n\}$.

This procedure lasts until $y_j = 0$, which means that there are no bracket codes less than y_{j-1} and that there are exactly $j-1$ different types of brackets contained in string S. Then condition $j-1 \leq t$ indicates whether Step 1 was executed correctly. We formalize this idea in the code listing of Algorithm 6, assuming subroutine $\text{QMAX}(q(*, y_1), 1, n)$ to implement the quantum algorithm for maximum search [9].

Algorithm 6. Step 1 for solving $\text{DYCK}'_{k,n,t}$

$j \leftarrow 1$
$y_1 \leftarrow \text{QMAX}(q(*, 2T+1), 1, n)$
while $y_j \neq 0$ **do**
   if $j > t$ **then**
      return 0
   end if
   $j \leftarrow j+1$
   $y_j \leftarrow \text{QMAX}(q(*, y_{j-1}), 1, n)$
end while
return 1

Lemma 5. *The query complexity of Algorithm 6 is $O(t\sqrt{n}\log t)$, and the error probability is some constant $\varepsilon < 1$.*

Proof. The expected query complexity of $\text{QMAX}(q(*, y_{j-1}), 1, n)$ is $O(\sqrt{n})$ [9]. According to Markov's inequality, also the exact query complexity of $\text{QMAX}(q(*, y_{j-1}), 1, n)$ is $O(\sqrt{n})$. As the error probability of QMAX is some constant, repeating it $2 \log_2 t$ times results in the error probability $O(\frac{1}{t^2})$.   □

If $t = O(\log n^{0.5(k-1)})$, then the query complexity of Algorithm 6 (run at Step 1) won't exceed the complexity of Step 2, and the overall complexity of the algorithm will remain the same.

Theorem 4. *Algorithm 4 with Step 1 implemented by Algorithm 6, solves* $\text{DYCK}'_{k,n,t}$. *If* $t = O(\log n^{0.5(k-1)})$, *then the query complexity of this solution is* $O(\sqrt{n}(\log n)^{0.5k})$, *and the two-side bounded error probability is* $\varepsilon < 0.5$.

Proof. According to Lemma 5, the query complexity of Step 1 is $O(\sqrt{n} \log n^{0.5(k-1)} \log \log n) = O(\sqrt{n} \log n^{0.5k})$. Steps 2 and 3 are the same as in Algorithm 4, with complexities resp. $O(\sqrt{n} \log n^{0.5k})$ and $O(\sqrt{n} \log n^{0.5(k-1)})$ proven as for Theorem 1. Thus the overall query complexity is

$$O(\sqrt{n} \log n^{0.5k}) + O(\sqrt{n} \log n^{0.5k}) + O(\sqrt{n} \log n^{0.5(k-1)}) = O(\sqrt{n} \log n^{0.5k}).$$

The estimation of the error probability is analogous to the one in the proof of Theorem 1   □

We strongly believe that there exists a more efficient quantum algorithm for solving Step 1 of $\text{DYCK}'_{k,n,t}$, but for now we settle for the just proposed iterative maximum search, and leave better approaches for the future work.

4.2 Property Testing

Researchers have an interest in recognizing the Dyck language (solving the problem DYCK) in the sense of property testing [20]. Imagine that we have a promise, that either an input S belongs to the language, or the Hamming distance from S to the language (i.e. to the closest word of the language) is at least $\varepsilon \cdot n$, where $\varepsilon < 1$ is some positive constant and n is the length of an input. It remains an open question, whether quantum computing can bring any speed-up for different values of ε.

4.3 Tighter Bounds

This paper along with [4] demonstrates the upper bound $O\left(\sqrt{n} \cdot \left(\sqrt{\log n}\right)^k\right)$ (Theorem 1) for the quantum query complexity and the lower bound $\Omega\left(\sqrt{n} \cdot c^k\right)$ (Theorem 2) for some constant c. There is a certain gap between these two bounds, and reducing this gap is another open problem. Note that similar gap also holds for Dyck language with one type of brackets [4].

Acknowledgments. This work was funded by the subsidy allocated to Kazan Federal University for the state assignment in the sphere of scientific activities, project No. 0671-2020-0065; supported by the Kazan Federal University Strategic Academic Leadership Program; partially funded by RFBR according to the research project No. 20-37-70080; supported by ERDF within the project "Quantum algorithms: from complexity theory to experiment" funded under programme 1.1.1.5.

References

1. Aaronson, S., Grier, D., Schaeffer, L.: A quantum query complexity trichotomy for regular languages. In: 2019 IEEE 60th Annual Symposium on Foundations of Computer Science (FOCS), pp. 942–965. IEEE (2019)
2. Ablayev, F., Ablayev, M., Huang, J.Z., Khadiev, K., Salikhova, N., Wu, D.: On quantum methods for machine learning problems part I: quantum tools. Big Data Mining Anal. **3**(1), 41–55 (2019)
3. Ambainis, A.: Understanding quantum algorithms via query complexity. In: Proceedings of the International Conference of Mathematics 2018, vol. 4, pp. 3283–3304 (2018)
4. Ambainis, A., et al.: Quantum lower and upper bounds for 2D-grid and Dyck language. In: 45th International Symposium on Mathematical Foundations of Computer Science (MFCS 2020), volume 170 of Leibniz International Proceedings in Informatics (LIPIcs), pp. 8:1–8:14 (2020)
5. Barr, K., Fleming, T., Kendon, V.: Simulation methods for quantum walks on graphs applied to formal language recognition. Nat. Comput. **14**(1), 145–156 (2014). https://doi.org/10.1007/s11047-014-9441-x
6. Boyer, M., Brassard, G., Høyer, P., Tapp, A.: Tight bounds on quantum searching. Fortschr. Phys. **46**(4–5), 493–505 (1998)
7. Brassard, G., Høyer, P., Mosca, M., Tapp, A.: Quantum amplitude amplification and estimation. Contemp. Math. **305**, 53–74 (2002)
8. Buhrman, H., Patro, S., Speelman, F.: A framework of quantum strong exponential-time hypotheses. In: 38th International Symposium on Theoretical Aspects of Computer Science (STACS 2021), volume 187 of Leibniz International Proceedings in Informatics (LIPIcs), pp. 19:1–19:19 (2021)
9. Dürr, C., Høyer, P.: A quantum algorithm for finding the minimum. arXiv:quant-ph/9607014 (1996)
10. Glos, A., Nahimovs, N., Balakirev, K., Khadiev, K.: Upper bounds on the probability of finding marked connected components using quantum walks. Quantum Inf. Process. **20**(1), 1–23 (2021)
11. Grover, L.K.: A fast quantum mechanical algorithm for database search. In: Proceedings of the Twenty-Eighth Annual ACM Symposium on Theory of Computing, pp. 212–219. ACM (1996)
12. Høyer, P., Mosca, M., de Wolf, R.: Quantum search on bounded-error inputs. In: Baeten, J.C.M., Lenstra, J.K., Parrow, J., Woeginger, G.J. (eds.) ICALP 2003. LNCS, vol. 2719, pp. 291–299. Springer, Heidelberg (2003). https://doi.org/10.1007/3-540-45061-0_25
13. Jordan, S.: Quantum algorithms zoo (2021). http://quantumalgorithmzoo.org/
14. Khadiev, K., Ilikaev, A.: Quantum algorithms for the most frequently string search, intersection of two string sequences and sorting of strings problems. In: International Conference on Theory and Practice of Natural Computing, pp. 234–245 (2019)
15. Kravchenko, D., Khadiev, K., Serov, D.: On the quantum and classical complexity of solving subtraction games. In: van Bevern, R., Kucherov, G. (eds.) CSR 2019. LNCS, vol. 11532, pp. 228–236. Springer, Cham (2019). https://doi.org/10.1007/978-3-030-19955-5_20
16. Khadiev, K., Safina, L.: Quantum algorithm for dynamic programming approach for DAGs. Applications for Zhegalkin polynomial evaluation and some problems on DAGs. In: McQuillan, I., Seki, S. (eds.) UCNC 2019. LNCS, vol. 11493, pp. 150–163. Springer, Cham (2019). https://doi.org/10.1007/978-3-030-19311-9_13

17. Khadiev, K., Mannapov, I., Safina, L.: The quantum version of classification decision tree constructing algorithm c5. 0. CEUR Workshop Proceedings 2500 (2019)

18. Kravchenko, D., Khadiev, K., Serov, D., Kapralov, R.: Quantum-over-classical advantage in solving multiplayer games. In: Schmitz, S., Potapov, I. (eds.) RP 2020. LNCS, vol. 12448, pp. 83–98. Springer, Cham (2020). https://doi.org/10.1007/978-3-030-61739-4_6

19. Nielsen, M.A., Chuang, I.L.: Quantum Computation and Quantum Information. Cambridge University Press, New York (2010)

20. Parnas, M., Ron, D., Rubinfeld, R.: Testing membership in parenthesis languages. Random Struc. Alg. **22**(1), 98–138 (2003)

Affine Automata Verifiers

Aliya Khadieva[1,2]([✉]) and Abuzer Yakaryılmaz[1,3] [ID]

[1] University of Latvia, Rīga, Latvia
`abuzer@lu.lv`
[2] Kazan Federal University, Kazan, Russia
[3] QWorld Association, Tallinn, Estonia
`https://qworld.net`

Abstract. We initiate the study of the verification power of Affine finite automata (AfA) as a part of Arthur-Merlin (AM) proof systems. We show that every unary language is verified by a real-valued AfA verifier. Then, we focus on the verifiers restricted to have only integer-valued or rational-valued transitions. We observe that rational-valued verifiers can be simulated by integer-valued verifiers, and their protocols can be simulated in nondeterministic polynomial time. We show that this upper bound is tight by presenting an AfA verifier for NP-complete problem SUBSETSUM. We also show that AfAs can verify certain non-affine and non-stochastic unary languages.

Keywords: Affine automata · Interactive proof systems · Arthur-Merlin games · Unary languages · Subset-sum problem · NP

1 Introduction

Affine finite automata (AfAs) are quantum-like generalization of probabilistic finite automata (PFAs) mimicking quantum interference and having the capability of "making measurement" based on ℓ_1-norm (called weighting). The computation of an AfA is linear, but the weighting operators may be non-linear.

AfAs was formally defined in [7], and it was shown that they are more powerful than PFAs and quantum finite automata (QFAs) in bounded error and unbounded error settings, but their nondeterministic version is equivalent to nondeterministic QFAs. Since then, AfAs and their different generalizations (e.g., OBDDs and counter automata) have been investigated in a series of work [14–16,18,24,31,32,34].

In this paper, we consider AfAs as part of Arthur-Merlin (AM) proof systems and investigate their verification power. We show that every unary language can be verified by a real-valued AfA verifier. Then, we focus on the verifiers with integer-valued or rational-valued transitions. We show how to simulate

arXiv:2104.11192.

I. Kostitsyna and P. Orponen (Eds.): UCNC 2021, LNCS 12984, pp. 84–100, 2021.
https://doi.org/10.1007/978-3-030-87993-8_6

rational-valued verifiers by integer-valued ones and how to simulate their protocols in nondeterministic polynomial time. We present an AfA verifier for NP-complete problem SUBSETSUM. We also show that AfAs can verify certain non-affine and non-stochastic unary languages. In our protocols, we use similar verification strategies and encoding techniques previously used for two-way QFAs in [29,33,37].

In the rest of this section, we review the previously known results on probabilistic, quantum, and affine automata to give a snapshot of literature and so to see where our new results are placed. The notations and definitions are given in Sect. 2, our main result on unary languages is given in Sect. 3, and our results on rational- or integer-valued AfAs are given in Sect. 4. We close the paper with a summary. The omitted parts due to space constraints are available in [20].

1.1 The Computational Power of AfAs Compared to PFAs and QFAs

We review the previously known results comparing AfAs with PFAs and QFAs.

The bounded error PFAs and QFAs recognize all and only regular languages [2,22,23,26]. But bounded error AfAs can recognize some nonregular languages such as $\text{UPAL} = \{a^n b^n \mid n > 0\}$ and $\text{PAL} = \{w \in \{a,b\}^* \mid w = w^r\}$ [7]. Moreover, AfAs can be very succinct compared to PFAs and QFAs [17,31,32,34], e.g., they recognize a family of regular languages with bounded error by using only two states, but the number of states of bounded error PFAs or QFAs cannot be bounded for this family.

The class of languages recognized by PFAs with cutpoints is called stochastic languages [26]. QFAs recognize all and only stochastic languages with cutpoints [35,38]. Similar to bounded error case, AfAs are more powerful than both, and they can recognize some nonstochastic languages [7]. On the other hand, in the nondeterministic setting (when the cutpoint is fixed to zero), QFAs and AfAs have the same computational power [7].

Regarding the limitations on the computational power of AfAs, we know that [16,30,32,36]:

- (one-sided or two-sided) bounded error rational-valued and integer-valued AfAs have the same computational power;
- one-sided bounded error rational-valued AfAs cannot recognize any nonregular unary language;
- algebraic-valued AfAs cannot recognize certain non-stochastic unary languages in L even with unbounded error (with cutpoints); and,
- the class of languages recognized by bounded error rational-valued AfAs is a proper subset of L.

Here, L is a class of languages recognizable in logarithmic space.

One open problem is whether two-sided bounded error rational-valued AfAs can recognize any nonregular unary language, and an untouched direction is the computational capabilities of real-valued AfAs (i.e., all known results have been given for rational- or interger-valued AfAs).

1.2   The Verification Power of PFAs and QFAs

Interactive proof systems (IPSs) [12] with PFA verifiers [9] can verify some non-regular languages such as TWIN = $\{wcw \mid w \in \{a,b\}^*\}$ with bounded error [28]. The same result is valid for IPS with QFA verifiers communicating with the prover classically[1]. IPSs are also called private-coin systems since a verifier can hide its probabilistic decisions from the prover. In this way, the verifier can use stronger verification strategies as a part of the protocol (between the verifier and prover) since the prover may not guess the exact configuration of the verifier, and so it may not easily mislead the verifier when it is not honest.

When the computation of a verifier is fully seen to the prover, the system is called public-coin IPS or AM system [3,5]. AM system with PFA verifiers [6] cannot recognize any nonregular languages with bounded error, and we do not know whether AM systems with QFA verifiers can recognize any nonregular language with bounded error.

When considering the known results for AfAs (Sect. 1.1), there are two natural questions about the verification power of AM systems with rational-valued AfA verifiers:

1. whether we can go beyond L and, if so, how far, and,
2. whether some nonregular unary languages can be verified or not.

We answer both questions positively, and we obtain NP as the tight upper bound for non-unary languages.

1.3   Two-Way PFAs and QFAs

As mentioned above, AfAs can recognize nonregular languages UPAL and PAL with bounded error without interacting with any prover. Similar results can be obtained for PFAs and QFAs when reading the input many times by using a two-way head [1,10]. We review basic facts about bounded error two-way PFAs and QFAs to have a better picture for our results on AfAs.

The language UPAL is recognized by bounded error two-way QFAs [1] in polynomial expected time and as well as by two-way PFAs [10] but only in exponential expected time [13].

The language PAL can be recognized by bounded error two-way QFAs in exponential expected time [1], but it cannot be recognized in polynomial expected time even if two-way QFAs are augmented with logarithmic amount of space [27]. On the other hand, AM systems with two-way PFA verifiers cannot verify PAL with bounded error even if augmented with logarithmic space [9]. Besides, two-way bounded error PFAs can recognize only regular languages in polynomial expected time [8], and it is open whether AM systems with two-way PFAs can verify any nonregular languages in polynomial time.

---

[1] When the proof system is fully quantum, we know little [25]: the restricted QFA model defined in [22] can verify only regular languages with bounded error.

Regarding unary languages, bounded error two-way PFAs cannot recognize any nonregular language [19], and it is open whether any unary nonregular language is verified by a bounded error AM system with two-way PFA verifier [6]. It is also open whether bounded error two-way QFAs can recognize any nonregular unary language.

The class of languages verified by AM systems with two-way rational-valued PFA verifiers is a proper subset of P [5]. Therefore, the verification power of AfAs can go beyond the verification power of two-way PFAs.

On the other hand, AM systems with two-way QFAs are very powerful [29,33]. Two-way QFAs can verify every unary language in exponential expected time, and so their verification power is equivalent to that of AfAs on unary languages. On non-unary languages, rational-valued two-way QFAs can verify every language in PSPACE and some NEXP-complete languages. Therefore, AM systems with rational-valued AfAs are weaker than AM systems with rational-valued two-way QFAs. Here we remark that AfA verifiers read the input once, but two-way QFAs may run in exponential or double-exponential expected time.

2   Preliminaries

Throughout the paper, $|\cdot|$ refers to the ℓ_1-norm; Σ denotes the input alphabet not containing symbols ¢ and \$, respectively called the left and right end-markers; $\tilde{\Sigma}$ is the set $\Sigma \cup \{¢,\$\}$; Σ^* denotes the set of all strings defined on the alphabet Σ including the empty string denoted ε; and, for a given string $w \in \Sigma^*$, $\tilde{w}$ denotes the string $¢w\$$. Moreover, for any string w, $|w|$ is the length of w, $|w|_\sigma$ is the number of occurrences of symbol σ in w, and, whenever $|w| > 0$, w_i represents the i-th symbol of w, where $1 \leq i \leq |w|$. For an automaton M, $f_M(w)$ is the accepting probability of M on the input $w \in \Sigma^*$.

A *realtime* automaton reads the given input symbol by symbol and from left to right. On each symbol, a realtime automaton can stay a fixed amount of steps. If there is no waiting steps, then it is called *strict realtime*. In this paper, we focus only on the strict realtime models. For every given input w, it is fed as $\tilde{w}$ so that the automaton can make pre-processing and post-processing while reading symbols ¢ and \$, respectively.

An m-state affine system is represented by $\mathbb{R}^m$, and affine state of this system is represented by an m-dimensional vector: $v = (\alpha_1 \ \cdots \ \alpha_m)^{\mathsf{T}} \in \mathbb{R}^m$ satisfying that $\sum_{j=1}^m \alpha_j = 1$, where α_j, similar to the amplitudes in quantum systems, is the value of the system being in state e_j.

We transform the affine system by applying affine operators to the vector v. Any affine operator of this system is a linear operator represented by an $(m \times m)$-dimensional matrix:

$$A = \begin{pmatrix} a_{1,1} & \cdots & a_{1,m} \\ \vdots & \ddots & \vdots \\ a_{m,1} & \cdots & a_{m,m} \end{pmatrix} \in \mathbb{R}^{m \times m}$$

satisfying that $\sum_{j=1}^{m} a_{j,i} = 1$ for each column i (the column summation is 1). When the operator A is applied to the affine state v, the new state is $v' = A \cdot v$.

To retrieve information from the affine system, similar to the measurement operators of quantum system, we apply a weighting operator. When the affine state v is weighted, the i-th state is observed with probability

$$\frac{|\alpha_i|}{|v|} = \frac{|\alpha_i|}{|\alpha_1| + \cdots + |\alpha_m|}.$$

If the system is restricted to have only the nonnegative real numbers, then it turns out to be a probabilistic system.

2.1 Finite Automata with Deterministic and Affine States

Similar to finite automata with quantum and classical states (QCFA) [1], a finite automaton with classical and affine states (ADfA) is an n-state deterministic finite automaton having an m-state affine register, where $m, n > 0$. (Even though both affine and classical parts have finite states, we use lowercase "f" to emphasize the nonlinearity of affine part.) Let $S = \{s_1, \ldots, s_n\}$ be the classical deterministic states and let $E = \{e_1, \ldots, e_m\}$ be the affine states, where e_i is the standard basis in $\mathbb{R}^m$ (all zeros but the i-th entry which is 1).

The computation is governed classically. During the computation, each transition of an ADfA has two parts: affine and classical parts.

1. Affine transition: For each pair of deterministic state and reading symbol, say (s, σ), either an affine operator or a weighting operator is applied to the affine register.
2. Classical transition can be two types:
   (a) If an affine operator is applied, then the next classical state is determined based on (s, σ).
   (b) If a weighting operator is applied, then the next classical state is determined based on (s, σ, e), where $e \in E$ is the measured affine state.

In this paper, we apply the weighting operator only after reading the whole input, and so, we keep the formal definition of the models simpler: a single transition updates both the classical and affine parts at the same time.

Formally, a ADfA M with n classical deterministic and m affine states is a 8-tuple

$$M = (S, E, \Sigma, \delta, s_I, e_I, s_a, E_a),$$

where

- S and E are the sets of states as specified above;
- δ is the transition function described below;
- $s_I \in S$ and $e_I \in E$ are the classical deterministic and affine initial states, respectively; and,
- $s_a \in S$ is the deterministic accepting state;

– $E_a \subseteq E$ is the set of affine accepting state(s).

Let $w \in \Sigma^*$ be the given input of length l. The ADfA reads the input as $\tilde{w} = \text{¢}w\$$ from left to right and symbol by symbol. The computation of M is traced by a pair (s, v) called a configuration, where $s \in S$ is the classic state, $v \in \mathbb{R}^m$ is a vector of the affine configuration. At the beginning of the computation, M is in (s_I, v_0), where the affine state $v_0 = e_I$.

The transition function is defined as $\delta : S \times \tilde{\Sigma} \to S \times \mathbb{R}^{m \times m}$. Let (s, v_j) be the configuration of M after the j-th step and $\sigma = \tilde{w}_{j+1} \in \tilde{\Sigma}$. Then the new configuration is (s', v_{j+1}), where $\delta(s, \sigma) = (s', A)$ and $v_{j+1} = Av_j$.

After reading \$ symbol, if the final classical state is not s_a, then the input is rejected deterministically: $f_M(w) = 0$. Otherwise, a weighting operator is applied. The input is accepted if an affine accepting state is observed. We denote the final state as $v_f = v_{|\tilde{w}|}$. Then, the accepting probability by the affine part is

$$f_M(w) = \frac{\sum_{e_i \in E_a} |v_f[i]|}{|v_f|} \in [0, 1].$$

We remark that the ADfA M defined here can be exactly simulated by the original model defined in [7] with $(m \cdot n)$ affine states.

2.2 Affine Automata Verifiers

In this paper, we study only Arthur-Merlin type of interactive proof systems where the verifiers are affine automata. In [6], Arthur-Merlin systems with probabilistic finite automata verifier are defined as an automata having both nondeterministic and probabilistic states. We follow the same framework here. We indeed give the ability of making nondeterministic transitions to ADfA models. The reason of this admission is that the prover of AM system has an unrestricted computational power and the whole access to the input. Thus, we can describe the AM protocol as providing a certificate in a nondeterministic automaton.

A finite automaton with nondeterministic and affine states (ANfA) with n classic nondeterministic and m affine states is formally an 8-tuple

$$N = (S, E, \Sigma, \delta, s_I, e_I, s_a, E_a),$$

where all elements are the same as ADfA except the transition function. For the pair $(s, \sigma) \in S \times \tilde{\Sigma}$, it can have one or more transitions:

$$\delta(s, \sigma) \to \{(s'_1, A_1), \dots, (s'_k, A_k)\},$$

where each pair (s, σ) can have a different $k > 0$ value. When having more than one transition, N picks each of them nondeterministically by creating a new path. In this way, N forms a computation tree, where the root is the starting configuration. Remark that the computation in each path is the same as that of ADfAs and each path may have a different accepting probability. Each path here refers to a different certificate.

2.3   Language Recognition/Verification and Classes

A language $L \subseteq \Sigma^*$ is said to be recognized by an ADfA M with error bound $\epsilon < \frac{1}{2}$, if (i) for every $w \in L$, $f_M(w) \geq 1 - \epsilon$, and (ii) for every $w \notin L$, $f_M(w) \leq \epsilon$. Shortly, we can also say that L is recognized by M with bounded error or L is recognized by a bounded error ADfA.

A language $L \subseteq \Sigma^*$ is said to be verified by an ANfA V with error bound $\epsilon < \frac{1}{2}$, if (i) for every $w \in L$, there is a path on which $f_V(w) \geq 1 - \epsilon$, and (ii) for every $w \notin L$, $f_V(w) \leq \epsilon$ on each path. Shortly, we can also say that L is verified by V with bounded error or L is verified by a bounded error ANfA.

We define AM(AfA) as the class of languages verifiable by bounded error Arthur-Merlin system having realtime affine finite verifiers. Any language verifiable by a bounded error ANfA is in this class, and we obtain all results in this paper by ANfAs. Remark that the model of realtime affine finite verifiers is more general model than ANfA as we can apply weighting operators more than once and also process the outcomes classically.

If the verifier is a PFA, QFA, two-way PFA, or two-way QFA, then the related class is AM(PFA), AM(QFA), AM(2PFA), or AM(2QCFA), respectively, where 2QCFA is the two-way QFA model defined in [1].

The classes $\mathsf{AM_Q}(\cdot)$ or $\mathsf{AM_Z}(\cdot)$ denote the AM classes where the verifiers are restricted to have rational-valued or integer-valued components, respectively.

Here is the list of standard complexity classes mentioned in the paper:

REG   : regular languages
L   : logarithmic space
P   : polynomial time
NP   : nondeterministic polynomial time
SPACE(n) : linear space
PSPACE   : polynomial space
NEXP   : nondeterministic exponential space

Lastly, for a given complexity class C, UnaryC denotes its unary version, and as a special case, the set of all unary languages is called TALLY.

3   Verification of Every Unary Language

Let $L \subseteq \Sigma^*$ be an arbitrary unary language, where $\Sigma = \{a\}$. We define a real number to encode the whole membership information of L as follows:

$$\alpha_L = \sum_{i=0}^{\infty} \frac{b_i}{32^{i+1}} = \frac{b_0}{32} + \frac{b_1}{32^2} + \frac{b_2}{32^3} + \cdots,$$

where $b_i = 1$ if $a^i \in L$ and $b_i = 0$ if $a^i \notin L$. In binary form: $bin(\alpha_L) = 0.0000b_00000b_1 \cdots 0000b_i \cdots$. Moreover, we define

$$\alpha_L[j] = \frac{b_j}{32} + \frac{b_{j+1}}{32^2} + \frac{b_{j+2}}{32^3} + \cdots, \text{ where } j \geq 0.$$

We give a few basic facts about α_L and $\alpha_L[j]$, which we will use in our proofs.

1. For any $\alpha_L[j]$, there is a unary language L' such that $\alpha_L[j] = \alpha_{L'}$.
2. The values of α_L and so $\alpha_L[j]$ are bounded:

$$0 \leq \alpha_L \leq \frac{1}{31} \text{ and } 0 \leq \alpha_L[j] \leq \frac{1}{31}.$$

3. The values of $\alpha_L[j+1]$ and $\alpha_L[j]$ can be related:
   - If $b_j = 0$: $\alpha_L[j+1] = 32 \cdot \alpha_L[j]$.
   - If $b_j = 1$: $\alpha_L[j+1] = 32 \cdot \alpha_L[j] - 1$.

By using α_L, we design a bounded error ANfA for language L. The main idea behind the protocol is that each b_i is nondeterministically guessed and the verification is done by subtracting the guessed b_i and the actual value b_i encoded in α_L. As long as the nondeterministic choices are correct, the result of such subtractions will be zero. Otherwise, it will not be zero, based on which we reject the input. The details are given in the proof below.

Theorem 1. *Every unary language $L \subseteq \{a\}^*$ is verified by an ANfA V with error bound 0.155.*

Proof. The verifier V has two classical states and three affine states, where s_2 is the classical accepting state and e_1 is the only affine accepting state. The initial affine state is $v_0 = (1 \ \ 0 \ \ 0)^\mathsf{T}$ and initial classical state is s_1.

Let $w = a^l$ be the given input $(l \geq 0)$. Until reading \$, V makes two non-deterministic transitions for each $\tilde{w}_i$ $(i \in \{1, \ldots, l+1\})$: V guesses the value of b_{i-1}, say g_{i-1}. If $g_{i-1} = 0$, then classical state is set to s_1, and if $g_{i-1} = 1$, then classical state is set to s_2. The affine operators are described below.

On symbol ¢, a combination of two affine operators is applied. In the first part, the affine state is set as

$$\begin{pmatrix} 1 \\ \alpha_L \\ -\alpha_L \end{pmatrix} = \begin{pmatrix} 1 & 0 & 0 \\ \alpha_L & 1 & 0 \\ -\alpha_L & 0 & 1 \end{pmatrix} \begin{pmatrix} 1 \\ 0 \\ 0 \end{pmatrix}.$$

In the second part, the affine operator A_{g_0} is applied, where

$$A_0 = \begin{pmatrix} 1 & -31 & -31 \\ 0 & 32 & 0 \\ 0 & 0 & 32 \end{pmatrix} \text{ and } A_1 = \begin{pmatrix} 1 & -31 & -31 \\ -1 & 32 & 0 \\ 1 & 0 & 32 \end{pmatrix}.$$

On each symbol a, the second part for symbol ¢ is repeated: the affine operator A_{g_i} is applied on the path where g_i is picked.

If b_0 is guessed correctly, then affine state becomes

$$\begin{pmatrix} 1 \\ \alpha_L[1] \\ -\alpha_L[1] \end{pmatrix} = A_{b_0} \begin{pmatrix} 1 \\ \alpha_L[0] \\ -\alpha_L[0] \end{pmatrix}.$$

It is sufficient to check the value of the second entry of v. Let us denote it $v[2]$.

- If $b_0 = 0$, after applying A_0, the value of $v[2]$ becomes $32 \cdot \alpha_L[0]$, which is equal to $\alpha_L[1]$.
- If $b_0 = 1$, after applying A_1, the value of $v[2]$ becomes $32 \cdot \alpha_L[0] - 1$, which is equal to $\alpha_L[1]$.

Similarly, as long as the nondeterministic guesses are correct, the affine part evolves as given below:

$$\begin{pmatrix} 1 \\ \alpha_L[1] \\ -\alpha_L[1] \end{pmatrix} \xrightarrow{1^{st}\ a} \begin{pmatrix} 1 \\ \alpha_L[2] \\ -\alpha_L[2] \end{pmatrix} \xrightarrow{2^{nd}\ a} \cdots \xrightarrow{l^{th}\ a} \begin{pmatrix} 1 \\ \alpha_L[l+1] \\ -\alpha_L[l+1] \end{pmatrix}.$$

Now, we examine the case in which at least one nondeterministic guess is wrong. Assume that $g_i \neq b_i$ is the first wrong guess (for symbol $\tilde{w}_{i+1}$). The value of $v[2]$ is $\alpha_L[i]$ before this guess, and it becomes

$$1 + \alpha_L[i+1] \quad \text{or} \quad \alpha_L[i+1] - 1$$

after the guess. Thus, the absolute value of $v[2]$ is bounded below by $1 - \frac{1}{31} = \frac{30}{31}$, which is at least 30 times greater than any $\alpha_L[j]$. If there is another symbol a to be read, then the value of $v[2]$ is multiplied by 32 followed by subtraction of 0 or -1. That means the integer part of the absolute of new value of $v[2]$ becomes greater than 30, and so the absolute value of $v[2]$ is at least 900 times greater than any $\alpha_L[j]$. For each new symbol of a, this factor (i.e., 30 and 900) will be multiplied by at least 30.

On symbol \$, V does not change the classical state and applies the following operator to the affine state:

$$A_\$(k) = \begin{pmatrix} 1 & 1-k & 1-k \\ 0 & k & 0 \\ 0 & 0 & k \end{pmatrix},$$

where $k = \frac{31}{2\sqrt{30}}$, which gives the minimum error when maximizing the accepting probability for members and minimizing the same for the non-members.

If $w \in L$, the path following the correct nondeterministic guesses ends in classical state s_2 and affine state $(1 \quad k \cdot \alpha_L[l+1] \quad -k \cdot \alpha_L[l+1])^\mathsf{T}$. Remember that $0 \leq \alpha_L[l+1] \leq \frac{1}{31}$. Thus, the input is accepted with probability

$$\frac{1}{1 + 2k\alpha_L[l+1]} \geq \frac{1}{1 + \frac{2k}{31}} = \frac{1}{1 + \frac{1}{\sqrt{30}}} = \frac{\sqrt{30}}{1 + \sqrt{30}} = 1 - \frac{1}{1 + \sqrt{30}} > 0.845.$$

If $w \notin L$, then we have different cases. (1) If b_l is guessed correctly ($g_l = 0$), then the input is rejected deterministically. (2) If each guess is correct except b_l ($g_l = 1$), then affine state is

$$\begin{pmatrix} 1 \\ k(\alpha_L[l+1] - 1) \\ -k(\alpha_L[l+1] - 1) \end{pmatrix},$$

and so, the accepting probability is

$$\frac{1}{1 + 2k(1 - \alpha_L[l+1])} \le \frac{1}{1 + 2k(\frac{30}{31})} = \frac{1}{1 + \sqrt{30}} < 0.155.$$

In other words, the rejecting probability is at least $1 - 0.155 = 0.845$. (3) If the guess g_i for $i < l$ is wrong, then, as we described above, the absolute values of $v[2]$ and $v[3]$ are at least 30 times bigger than that of the case (2), and so is the rejecting probability.   □

When defining α_L, the denominators can be some numbers greater than 32, and, in this way we can obtain better error bounds, i.e., arbitrarily close to 0.

Corollary 1. *Every unary language $L \subseteq \{a\}^*$ is verified by ANfAs with arbitrarily small error bounds.*

4 $AM_{\mathbb{Z}}(AfA)$

Recently, it was shown [16] that any language recognized by a rational-valued ADfA with error bound ϵ is recognized by an integer-valued ADfA with error bound ϵ', where $0 \le \epsilon \le \epsilon' < \frac{1}{2}$. The latter automaton is constructed by modifying the components of the former automaton so that, on the same input, the accepting probability of the latter one can differ insignificantly from the accepting probability of the former one, i.e., the difference is at most $\epsilon' - \epsilon$. Thus, if the automaton is an ANfA, then on the same input, the accepting probabilities for the same nondeterministic path will differ insignificantly, and so the error bound increases but still less than $\frac{1}{2}$.

Theorem 2. $AM_{\mathbb{Q}}(AfA) = AM_{\mathbb{Z}}(AfA)$.

It is known that $AM(PFA) = REG$ [6]. We do not know whether $AM(QFA)$ contains any non-regular language. On the other hand, ADfAs can recognize some non-regular ones with bounded error such as PAL requiring at least logarithmic space for bounded error probabilistic computation [11]. A natural question is whether $AM(AfA)$ goes beyond L.

Theorem 3. $AM_{\mathbb{Q}}(AfA) \subseteq NP \cap SPACE(n)$.

Proof. Let $L \in \mathsf{AM_Q(AfA)}$ be a language. Then, there is an affine automaton verifier V verifying L with error bound $\epsilon \in \mathbb{Q} \cap [0, \frac{1}{2})$.

The descriptions of V and the error bound are finite, which can be wired into the description of Turing Machines (TMs). For any given input, the computation on each path of V can be traced by vector and matrix multiplications. As the length of each sequence is linear, all computation including weighting, calculating the accepting probability, and comparing it with the error bound can be done in polynomial time and linear space (i.e., the size of affine state vector is fixed, the precision of each entry can be at most linear, and each new entry is a linear combination of these entries).

In the case of nondeterministic TM simulation, the TM implements the nondeterministic choices of V directly. In the case of linear-space TM simulation, the TM uses a linear counter to check all nondeterministic strategies one-by-one. Even though the overall simulation runs in exponential expected time, the space usage can be bounded linearly.   □

We show that integer-valued ANfAs can verify some NP-complete problems. We use the following language version of the Knapsack Problem (Page 491 of [21]): SUBSETSUM is the language of strings of the form $S\#B_1\# \cdots \#B_k$, where

- $S, B_1, \ldots, B_k \in \{0,1\}^*$ are binary numbers and
- there exists a subset of $\{B_1, \ldots, B_k\}$ that adds up to precisely S: that is, $\exists I \subseteq \{1, \ldots, k\}$ such that $S = \sum_{i \in I} B_i$.

Remark that we do not use any negative integer, and it is still NP-complete.

Theorem 4. SUBSETSUM *is verified by an integer-valued ANfA $V(t)$ such that every member is accepted with probability 1 and every non-member is accepted with probability at most $\frac{1}{2t+1}$ for some $t \in \mathbb{Z}^+$.*

Proof. Let $w \in \Sigma^*$, where $\Sigma = \{0, 1, \#\}$. The verifier $V(t)$, shortly V, classically checks that w has at least one $\#$. Otherwise, the input is rejected deterministically.

In the remaining part, we assume that w is of the form $S\#B_1\# \cdots \#B_k$ for some $k > 0$. Remark that the binary value of empty string is zero (whenever $S = \varepsilon$ or any $B_i = \varepsilon$). The protocol has the following steps:

1. V starts with encoding S into the value of affine state e_2.
2. V nodeterministically picks some B_i's ($1 \le i \le k$). Such decision is made when reading symbols $\#$.
   (a) If B_i is not picked, then affine state is not changed.
   (b) Otherwise, V encodes B_i into the value of the affine state e_3, and then, it is subtracted from the value of e_2 and the value of e_3 is set to zero.
3. At the end of the computation, the decision is made based on the fact that the value of e_2 is zero for the members in at least one path and non-zero integer for the non-members in each path. The error is decreased by using certain tricks before the weighting operator.

The affine part has four states $\{e_1, \ldots, e_4\}$ and e_1 is the only accepting state. The initial affine state is $(1\ 0\ 0\ 0)^\mathsf{T}$, and it does not change when reading ¢. For encoding binary string, we use the technique described in [20]. The value of S is encoded by using the affine operators $\{A_\sigma \mid \sigma \in \{0,1\}\}$:

$$
A_0 = \begin{pmatrix} 1 & 0 & 0 & 0 \\ 0 & 2 & 0 & 0 \\ 0 & 0 & 1 & 0 \\ 0 & -1 & 0 & 1 \end{pmatrix} \quad \text{and} \quad A_1 = \begin{pmatrix} 1 & 0 & 0 & 0 \\ 1 & 2 & 0 & 0 \\ 0 & 0 & 1 & 0 \\ -1 & -1 & 0 & 1 \end{pmatrix},
$$

where the value of e_3 is not changed. The value of each picked B_i is encoded by the affine operators $\{A'_\sigma \mid \sigma \in \{0,1\}\}$:

$$
A'_0 = \begin{pmatrix} 1 & 0 & 0 & 0 \\ 0 & 1 & 0 & 0 \\ 0 & 0 & 2 & 0 \\ 0 & 0 & -1 & 1 \end{pmatrix} \quad \text{and} \quad A'_1 = \begin{pmatrix} 1 & 0 & 0 & 0 \\ 0 & 1 & 0 & 0 \\ 1 & 0 & 2 & 0 \\ -1 & 0 & -1 & 1 \end{pmatrix},
$$

where the value of e_2 is not changed. With the following operator, the value of e_3 is subtracted from the value of e_2 and set to 0:

$$
D = \begin{pmatrix} 1 & 0 & 0 & 0 \\ 0 & 1 & -1 & 0 \\ 0 & 0 & 0 & 0 \\ 0 & 0 & 2 & 1 \end{pmatrix}.
$$

For a picked subset $I \subseteq \{1, \ldots, k\}$, let $S_I = \sum_{i \in I} B_i$. Before weighting operator, for some $t \in \mathbb{Z}^+$, we apply the following operator to decrease the error bound for the non-members:

$$
E(t) = \begin{pmatrix} 1 & 0 & 0 & 0 \\ 0 & t & 0 & 0 \\ 0 & 1-t & 1 & 1-t \\ 0 & 0 & 0 & t \end{pmatrix}.
$$

On the path where I is followed, just before applying $E(t)$, the affine state is

$$
\begin{pmatrix} 1 \\ S - S_I \\ 0 \\ S_I - S \end{pmatrix},
$$

which becomes

$$
\begin{pmatrix} 1 \\ t(S - S_I) \\ 0 \\ t(S_I - S) \end{pmatrix}
$$

after applying $E(t)$. It is easy to see that if $S = S_I$, then the final affine state is $(1 \ 0 \ 0 \ 0)^\mathsf{T}$ and so the input is accepted with probability 1. If $S \neq S_I$, then $|S - S_I| \in \mathbb{Z}^+$, and so the values of e_2 and e_4 are not zero and the accepting probability can be at most $\dfrac{1}{2t + 1}$.

Therefore, if $w \in \mathsf{SUBSETSUM}$, then there exists a subset I satisfying the membership condition and it is picked on a path where the input is accepted with probability 1. If $w \notin \mathsf{SUBSETSUM}$, there is no subset satisfying the membership condition, and so the input is accepted with probability at most $\dfrac{1}{2t + 1}$ in each path. The error bound can be arbitrarily small when $t \to \infty$.   □

It is not known whether there is an NP-complete unary language: if there is such a language, then P = NP [4]. Regarding the verification power of rational-valued ANfAs, we use some non-stochastic unary languages (unary languages that are not recognizable by any probabilistic finite automata with cutpoints).

For a given non-linear polynomial with nonnegative integer coefficients $P(x)$, the unary language $\mathsf{UPOLY(P)} = \{a^{P(i)} \mid i \in \mathbb{N}\}$ was shown to be not stochastic [30]. Recently, it was shown that [16] this family of languages cannot be recognized by algebraic-valued ADfAs even with cutpoints. We show that ANfAs can verify any $\mathsf{UPOLY(P)}$ language with bounded error.

We start with a simpler case $\mathsf{USQUARE}$, and then, based on it, we provide the proof for general case.

Theorem 5. *Language* $\mathsf{USQUARE}$ *is verified by an ANfA* $V(t)$ *with any error bound* $\frac{1}{2t+1}$, *where* $t \in \mathbb{Z}^+$.

Proof. We use the parameter t at the end of the proof, and we represent $V(t)$ shortly as V. The verifier V uses 4 affine states, and e_1 is the single accepting affine state. Let $w = 0^l$ be the given input. If $w = \varepsilon$, then it is accepted classically. We assume that $w \neq \varepsilon$ in the rest of the proof.

The protocol of V is as follows: V nondeterministically picks a positive integer $j \geq l$ and then checks whether $j^2 = l$. If $w \in \mathsf{USQUARE}$, then there exists such $j = \sqrt{l}$ and so this comparison is made successfully in one of the nondeterministic paths. If $w \notin \mathsf{USQUARE}$, there is no such j and so there is no successful comparison in any nondeterministic path.

The verifier follows $(l + 1)$ different paths during its computation:

$$path_0, path_1, \ldots, path_l,$$

where the main one is $path_0$. We use the encoding techniques described in [20]. When reading the i-th symbol of w, $path_0$ continues with $path_0$ or creates $path_i$.

On $path_0$, V is in the following affine states after reading w_i and w_l:

$$v_{0,i+1} = \begin{pmatrix} 1 \\ i \\ i^2 \\ 1 \end{pmatrix} \text{ and } v_{0,l+1} = \begin{pmatrix} 1 \\ l \\ l^2 \\ 1 \end{pmatrix},$$

respectively. After reading w_i, V creates $path_i$, on which it is in the affine state

$$v_{i,i+1} = \begin{pmatrix} 1 \\ i \\ i^2 \\ \overline{1} \end{pmatrix}.$$

For the rest of the computation, V continues with counting the number of symbols on $v[2]$, which is the second entry of the vector v, but it does not change the value of $v[3]$ until reading \$. The affine state on $path_i$ $(i > 0)$ after reading w_l is

$$v_{i,l+1} = \begin{pmatrix} 1 \\ l \\ i^2 \\ \overline{1} \end{pmatrix}.$$

On $path_0$, the input is rejected classically. On $path_i$, after reading \$, V enters the classical accepting state, and it sets the affine state as

$$\begin{pmatrix} 1 \\ t(l - i^2) \\ t(i^2 - l) \\ 0 \end{pmatrix}.$$

If $w \in L$, then on $path_{\sqrt{l}}$, the final affine state is e_1 and so w is accepted with probability 1.

If $w \notin L$, then on $path_i$, the absolute value of $v[2]$ or $v[3]$ is $|t(l - i^2)|$, which is at least t. Thus, the input is accepted with probability at most $\epsilon = \frac{1}{2t+1} \leq \frac{1}{3}$. It is clear that $\epsilon \to 0$ when $t \to \infty$. □

Theorem 6. *Language* UPOLY(P) *is verified by an ANfA* $V(t)$ *with any error bound* $\frac{1}{2t+1}$, *where* $t \in \mathbb{Z}^+$.

Proof. The proof is identical to the proof of Theorem 5 after modify the encoding part (we use the techniques described in [20]). First note that $P(i) \geq i$ since the coefficients of P are nonnegative. So, for any $0^l \in$ UPOLY(P), there exists $j \leq l$ such that $l = P(j)$. Second, on $path_i$, $P(i)$ is calculated and then the verifier checks whether $P(i) = l$ or not.

If 0^l is in UPOLY(P), then it is accepted with probability 1 in one of the nondeterministic paths. If it is not in UPOLY(P), then the accepting probability on any path can be at most $\frac{1}{2t+1}$. □

5   Summary

On unary languages, for the real-valued verifiers, we show that AfAs and 2QCFAs have the same verification power: TALLY = UnaryAM(2QCFA) = UnaryAM(AfA), where AfAs are realtime machines but 2QCFAs run in exponential expected time.

On unary languages, for the rational-valued verifiers, we know that

$$\mathsf{UnaryREG} = \mathsf{UnaryAM_{\mathbb{Q}}(PFA)} \subseteq \frac{\mathsf{UnaryAM_{\mathbb{Q}}(QFA)} \subseteq \mathsf{UnaryAM(QFA)}}{\mathsf{UnaryAM_{\mathbb{Q}}(2PFA)} \subseteq \mathsf{UnaryAM(2PFA)}},$$

where it is open if the inclusions are strict, and we show that $\mathsf{UPOLY(P)} \in \mathsf{UnaryAM_{\mathbb{Q}}(AfA)}$ and so we have $\mathsf{UnaryREG} \subsetneq \mathsf{UnaryAM_{\mathbb{Q}}(AfA)}$.

On non-unary languages, for the rational-valued verifiers, we give an upper bound for $\mathsf{AM_{\mathbb{Q}}(AfA)}$, and so we have

$$\mathsf{AM_{\mathbb{Q}}(AfA)} = \mathsf{AM_{\mathbb{Z}}(AfA)} \subseteq \mathsf{NP} \cap \mathsf{SPACE}(n) \subsetneq \mathsf{AM_{\mathbb{Q}}(2QCFA)},$$

where 2QCFAs run in double-exponential expected time. Our upper bound is tight since we show that $\mathsf{SUBSETSUM} \in \mathsf{AM_{\mathbb{Z}}(AfA)}$.

Acknowledgments. We thank to anonymous reviewers for their helpful comments. Yakaryılmaz was partially supported by the ERDF project Nr. 1.1.1.5/19/A/005 "Quantum computers with constant memory". A part of research is funded by the subsidy allocated to Kazan Federal University for the state assignment in the sphere of scientific activities, project No. 0671-2020-0065. A part of the research has been supported by the Kazan Federal University Strategic Academic Leadership Program. The research was partially funded by RFBR according to the research project No. 20-37-70080.

References

1. Ambainis, A., Watrous, J.: Two-way finite automata with quantum and classical states. Theor. Comput. Sci. **287**(1), 299–311 (2002)
2. Ambainis, A., Yakaryılmaz, A.: Automata: from mathematics to applications. Technical Report. arXiv:1507.01988 (2015). to appear in Automata and Quantum Computing edited by Jean-Éric Pin
3. Babai, L.: Trading group theory for randomness. In: STOC 1985, pp. 421–429 (1985)
4. Berman, P.: Relationship between density and deterministic complexity of MP-complete languages. In: Ausiello, G., Böhm, C. (eds.) ICALP 1978. LNCS, vol. 62, pp. 63–71. Springer, Heidelberg (1978). https://doi.org/10.1007/3-540-08860-1_6
5. Condon, A.: The complexity of space bounded interactive proof systems. In: Complexity Theory: Current Research, pp. 147–190. Cambridge University Press, Cambridge (1993)
6. Condon, A., Hellerstein, L., Pottle, S., Wigderson, A.: On the power of finite automata with both nondeterministic and probabilistic states. SIAM J. Comput. **27**(3), 739–762 (1998)
7. Díaz-Caro, A., Yakaryılmaz, A.: Affine computation and affine automaton. In: Kulikov, A.S., Woeginger, G.J. (eds.) CSR 2016. LNCS, vol. 9691, pp. 146–160. Springer, Cham (2016). https://doi.org/10.1007/978-3-319-34171-2_11, arXiv:1602.04732
8. Dwork, C., Stockmeyer, L.: A time complexity gap for two-way probabilistic finite-state automata. SIAM J. Comput. **19**(6), 1011–1123 (1990)

9. Dwork, C., Stockmeyer, L.: Finite state verifiers I: the power of interaction. J. ACM **39**(4), 800–828 (1992)

10. Freivalds, R.: Probabilistic two-way machines. In: Gruska, J., Chytil, M. (eds.) MFCS 1981. LNCS, vol. 118, pp. 33–45. Springer, Heidelberg (1981). https://doi.org/10.1007/3-540-10856-4_72

11. Freivalds, R., Karpinski, M.: Lower space bounds for randomized computation. In: Abiteboul, S., Shamir, E. (eds.) ICALP 1994. LNCS, vol. 820, pp. 580–592. Springer, Heidelberg (1994). https://doi.org/10.1007/3-540-58201-0_100

12. Goldwasser, S., Micali, S., Rackoff, C.: The knowledge complexity of interactive proof systems. SIAM J. Comput. **18**(1), 186–208 (1989)

13. Greenberg, A.G., Weiss, A.: A lower bound for probabilistic algorithms for finite state machines. J. Comput. Syst. Sci. **33**(1), 88–105 (1986)

14. Hirvensalo, M., Moutot, E., Yakaryılmaz, A.: On the computational power of affine automata. In: Drewes, F., Martín-Vide, C., Truthe, B. (eds.) LATA 2017. LNCS, vol. 10168, pp. 405–417. Springer, Cham (2017). https://doi.org/10.1007/978-3-319-53733-7_30

15. Hirvensalo, M., Moutot, E., Yakaryılmaz, A.: Computational limitations of affine automata. In: McQuillan, I., Seki, S. (eds.) UCNC 2019. LNCS, vol. 11493, pp. 108–121. Springer, Cham (2019). https://doi.org/10.1007/978-3-030-19311-9_10

16. Hirvensalo, M., Moutot, E., Yakaryılmaz, A.: Computational limitations of affine automata and generalized affine automata. Nat. Comput. **46**, 1–12 (2021). https://doi.org/10.1007/s11047-020-09815-1

17. Ibrahimov, R., Khadiev, K., Prūsis, K., Yakaryılmaz, A.: Error-free affine, unitary, and probabilistic OBDDs. Int. J. Found. Comput. Sci. https://doi.org/10.1142/S0129054121500246

18. Ibrahimov, R., Khadiev, K., Prūsis, K., Yakaryılmaz, A.: Error-free affine, unitary, and probabilistic OBDDs. In: Konstantinidis, S., Pighizzini, G. (eds.) DCFS 2018. LNCS, vol. 10952, pp. 175–187. Springer, Cham (2018). https://doi.org/10.1007/978-3-319-94631-3_15, arXiv:1703.07184

19. Kaņeps, J.: Regularity of one-letter languages acceptable by 2-way finite probabilistic automata. In: Budach, L. (ed.) FCT 1991. LNCS, vol. 529, pp. 287–296. Springer, Heidelberg (1991). https://doi.org/10.1007/3-540-54458-5_73

20. Khadieva, A., Yakaryılmaz, A.: Affine automata verifiers. Technical Report. arXiv:2104.11192 (2021)

21. Kleinberg, J., Tardos, É.: Algorithm Design. Pearson/Addison-Wesley, Boston (2006)

22. Kondacs, A., Watrous, J.: On the power of quantum finite state automata. In: FOCS 1997, pp. 66–75 (1997)

23. Li, L., Qiu, D., Zou, X., Li, L., Wu, L., Mateus, P.: Characterizations of one-way general quantum finite automata. Theor. Comput. Sci. **419**, 73–91 (2012)

24. Nakanishi, M., Khadiev, K., Prūsis, K., Vihrovs, J., Yakaryılmaz, A.: Exact affine counter automata. In: 15th International Conference on Automata and Formal Languages. EPTCS, vol. 252, pp. 205–218 (2017). arXiv:1703.04281

25. Nishimura, H., Yamakami, T.: An application of quantum finite automata to interactive proof systems. J. Comput. Syst. Sci. **75**(4), 255–269 (2009)

26. Rabin, M.O.: Probabilistic automata. Inf. Control **6**, 230–243 (1963)

27. Remscrim, Z.: Lower bounds on the running time of two-way quantum finite automata and sublogarithmic-space quantum turing machines. In: 12th Innovations in Theoretical Computer Science Conference. LIPIcs, vol. 185, pp. 39:1–39:20 (2021). https://doi.org/10.4230/LIPIcs.ITCS.2021.39

28. Say, A.C.C., Yakaryılmaz, A.: Finite state verifiers with constant randomness. Logical Methods Comput. Sci. **10**(3) (2014)
29. Say, A.C.C., Yakaryılmaz, A.: Magic coins are useful for small-space quantum machines. Quantum Inf. Comput. **17**(11 & 12), 1027–1043 (2017)
30. Turakainen, P.: On nonstochastic languages and homomorphic images of stochastic languages. Inf. Sci. **24**(3), 229–253 (1981)
31. Villagra, M., Yakaryılmaz, A.: Language recognition power and succinctness of affine automata. In: Amos, M., Condon, A. (eds.) UCNC 2016. LNCS, vol. 9726, pp. 116–129. Springer, Cham (2016). https://doi.org/10.1007/978-3-319-41312-9_10
32. Villagra, M., Yakaryılmaz, A.: Language recognition power and succinctness of affine automata. Nat. Comput. **17**(2), 283–293 (2017). https://doi.org/10.1007/s11047-017-9652-z
33. Yakaryılmaz, A.: Public qubits versus private coins. In: The Proceedings of Workshop on Quantum and Classical Complexity, pp. 45–60. University of Latvia Press (2013). eCCC:TR12-130
34. Yakaryılmaz, A.: Improved constructions for succinct affine automata. Technical Report. arXiv:2106.16197 (2021)
35. Yakaryılmaz, A., Say, A.C.C.: Languages recognized with unbounded error by quantum finite automata. In: Frid, A., Morozov, A., Rybalchenko, A., Wagner, K.W. (eds.) CSR 2009. LNCS, vol. 5675, pp. 356–367. Springer, Heidelberg (2009). https://doi.org/10.1007/978-3-642-03351-3_33
36. Yakaryılmaz, A., Say, A.C.C.: Languages recognized by nondeterministic quantum finite automata. Quantum Inf. Comput. **10**(9 & 10), 747–770 (2010)
37. Yakaryılmaz, A., Say, A.C.C.: Succinctness of two-way probabilistic and quantum finite automata. Discrete Math. Theor. Comput. Sci. **12**(2), 19–40 (2010)
38. Yakaryılmaz, A., Say, A.C.C.: Unbounded-error quantum computation with small space bounds. Inf. Comput. **279**(6), 873–892 (2011)

String Assembling Systems: Comparison to Sticker Systems and Decidability

Martin Kutrib$^{(\boxtimes)}$ and Matthias Wendlandt

Institut für Informatik, Universität Giessen, Arndtstr. 2, 35392 Giessen, Germany
{kutrib,matthias.wendlandt}@informatik.uni-giessen.de

Abstract. A string assembling system is a generative model that generates strings from copies out of a finite set of assembly units. The underlying mechanism is based on piecewise assembly of a double-stranded sequence of symbols, where the upper and lower strand have to match. So, the generative power of such systems is driven by the power of double-strands. Here we compare the generative capacity of string assembling systems with those of different variants of sticker systems. Though both types of systems seem to be closely related, we show that their generative capacities are different. In particular, it turns out that the family of languages generated by string assembling systems is incomparable with several language families induced by sticker systems.

Moreover, we consider decidability questions for the family of languages generated by string assembling systems and solve the remaining open case of systems whose assembly units are 2-length-restricted. It is shown that emptiness and several other questions are not even semi-decidable which improves the previously undecidability results furthermore.

1 Introduction

The advent of investigations of devices and operations that are inspired by the study of biological processes and the growing interest in nature-based problems modeled in formal systems, advises to examine the control mechanism of complementary double strands, which leads back to the definition of the Post Correspondence Problem [12].

Sticker systems are one approach to classify systems generating double strands. They were introduced in [4] in their basic variant. Afterwards different variants of sticker systems have been investigated [1,10,11]. A sticker system basically consists of dominoes that can be seen as double-stranded molecules and, thus, in connection with DNA it can be seen as a (possibly incomplete) part of a DNA strand. A main restriction of sticker systems is that the upper and the lower strand of the dominoes are glued together. Thus, if a double-stranded domino is assembled to a double strand, a basic restriction is that the length difference of the upper and the lower strand of the derived molecule fits to the length difference of the upper and the lower strand of the domino. This implies that applying only these kind of dominoes does not increase the

© Springer Nature Switzerland AG 2021
I. Kostitsyna and P. Orponen (Eds.): UCNC 2021, LNCS 12984, pp. 101–115, 2021.
https://doi.org/10.1007/978-3-030-87993-8_7

length difference of the two strands arbitrarily. Such so-called sticker systems with bounded delay are at most as powerful as linear context-free grammars. The property that dominoes with two strands cannot be chopped into two independent strands seems to be a big restriction, since the arbitrary growing of one strand increases the capacity of derivations. An inspiring definition is the Post Correspondence Problem, where the two strings may be applied independently to the appropriate sequences. There is a second property that limits the generative capacity. Consider a sticker system S and a derivation of S where both strands of a double-stranded sequence w have the same length. Then it is possible to prolong w by every double-stranded sequence that can be generated by S (possibly at two ends). There is no possibility to control, how and if the further computation goes on.

String assembling systems are another possibility to use double-stranded sequences to generate formal languages. They have been introduced in [7], where connections to one-way two-head finite automata are shown. As for sticker systems, also two-way variants have been considered [6]. The impact of the model-inherent control mechanisms is studied in [8].

In contrast to sticker systems, where dominoes are sticked together, the definition of string assembling systems (SAS) is basically motivated by the mechanisms of the Post Correspondence Problem. The assembly units are pairs of substrings that have to be connected to the upper and lower string generated so far synchronously. In comparison to sticker systems, the substrings are not glued together. This property enables the possibility to increase the length difference between the two strands arbitrarily and, moreover, compare positions that are an arbitrary long distance away from each other. Thus, it is possible to generate non-context-free languages even if the string assembling systems are defined to work one-way. Additionally, string assembling systems obey two further control mechanisms. First, it is required that the first symbol of a substring that is to be assembled has to match the last symbol of the strand to which it is connected. One can imagine that both symbols are glued together, one at the top of the other and, thus, just one appears in the final double strand. This property enables more possibilities to give languages structure in contrast to instances of the Post Correspondence Problem, where this property lacks. Second, it is distinguished between assembly units that may appear at the beginning, during, and at the end of the assembling process as for the notion of strictly locally testable languages [9,13].

The paper is organized as follows. In the next section we present preliminaries and introduce the systems we are interested in more formally. Examples are given and the hierarchy of language families induced by sticker systems is depicted in Fig. 2. Section 3 is devoted to compare the generative capacities of sticker systems to several variants of sticker systems. In Sect. 4 we obtain the non-semidecidability of emptiness, finiteness, infiniteness, equivalence, inclusion, regularity, and context-freeness for string assembling systems whose units are 2-length-restricted. Finally, we give some concluding remarks in Sect. 5.

2   Preliminaries and Definitions

We write Σ^* for the set of all words (strings) over the finite alphabet Σ. The empty word is denoted by λ, and $\Sigma^+ = \Sigma^* \setminus \{\lambda\}$. The reversal of a word w is denoted by w^R and for the length of w we write $|w|$. Generally, for a singleton set $\{a\}$ we also simply write a. We use $\subseteq$ for inclusions and $\subset$ for strict inclusions. In order to avoid technical overloading in writing, two languages L and L' are considered to be equal, if they differ at most by the empty word, that is, $L \setminus \{\lambda\} = L' \setminus \{\lambda\}$.

String Assembling Systems. We are especially interested in how string assembling systems can be used to describe languages. To this end, we consider arbitrary alphabets and do not restrict on the natural alphabet $\{A, G, C, T\}$. Clearly, there are ways to encode an arbitrary alphabet in the natural alphabet.

A string assembling system (SAS) generates a double string. For the generation, the basic element is a unit. A *unit* $u = (w_1, w_2)$ with $w_1, w_2 \in \Sigma^+$ consists of two strings over a given alphabet Σ. The first string w_1 is connected to the upper strand, while the second one w_2 is connected to the lower one. They can only be connected, (i) if the first symbol of w_1 is equal to the last symbol of the upper strand generated so far and, similarly, (ii) if the first symbol of w_2 is equal to the last symbol of the lower strand generated so far. In this case, the first symbols of the unit and the last symbols of the double stranded sequence generated so far are glued together on top of each other if the derived upper and the lower strands match at each position.

Each *generation* has to begin with a unit out of the set of *axioms*. Afterwards the derivation continues with units of a second set, the set of *assembly units*, and it has to be finished by an *ending unit*. The generation is said to be *valid* if and only if both strands are identical in each position when the derivation process stops.

Formally, a *string assembling system* (SAS) is a system $S = \langle \Sigma, A, T, E \rangle$, where Σ is a finite, nonempty set of *symbols*, $A \subset \Sigma^+ \times \Sigma^+$ is the finite set of *axioms* of the forms (uv, u) or (u, uv), where $u \in \Sigma^+$ and $v \in \Sigma^*$, $T \subset \Sigma^+ \times \Sigma^+$ is the finite set of *assembly units*, and $E \subset \Sigma^+ \times \Sigma^+$ is the finite set of *ending assembly units* of the forms (vu, u) or (u, vu), where $u \in \Sigma^+$ and $v \in \Sigma^*$.

The next definition formally says how the units are assembled.

Let $S = \langle \Sigma, A, T, E \rangle$ be an SAS. The *derivation relation* $\Rightarrow$ is defined on $\Sigma^+ \times \Sigma^+$ by

1. $(uv, u) \Rightarrow (uvx, uy)$ if
   i) $uv = ta$, $u = sb$, and $(ax, by) \in T \cup E$, for $a, b \in \Sigma$, $x, y, s, t \in \Sigma^*$, and
   ii) $vx = yz$ or $vxz = y$, for $z \in \Sigma^*$,
2. $(u, uv) \Rightarrow (uy, uvx)$ if
   i) $uv = ta$, $u = sb$, and $(by, ax) \in T \cup E$, for $a, b \in \Sigma$, $x, y, s, t \in \Sigma^*$, and
   ii) $vx = yz$ or $vxz = y$, for $z \in \Sigma^*$.

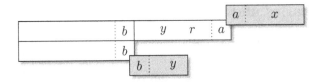

Fig. 1. Example of assembling a unit (ax, by) at the right.

An illustration can be found in Fig. 1.

A derivation is said to be *successful* if it initially starts with an axiom from A, continues with assembling units from T, and ends with assembling an ending unit from E. The process stops when an ending assembly unit is added. The sets A, T, and E are not necessarily disjoint.

The *language* $L(S)$ *generated* by S is defined to be the set

$$L(S) = \{ w \in \Sigma^+ \mid (u, v) \Rightarrow^* (w, w) \text{ is a successful derivation} \},$$

with $(u, v) \in A$, where $\Rightarrow^*$ refers to the reflexive, transitive closure of the derivation relation $\Rightarrow$.

Example 1. The SAS $S = \langle \{a, b, c\}, A, T, E \rangle$ generates the non-context-free language $\{ a^n b^n c^n \mid n \geq 1 \}$, where

$$A = \{(a, a)\}, \quad T = T_a \cup T_b \cup T_c, \quad E = \{(c, c)\},$$
$$T_a = \{(aa, a), (ab, a)\}, \quad T_b = \{(bb, aa), (bc, ab)\},$$
$$T_c = \{(cc, bb), (c, bc), (c, cc)\}.$$

The units in T_a are used to generate the prefixes $a^n b$. Initially, only the unit (aa, a) is applicable repeatedly. Then only (ab, a) can be used to generate the upper string $a^n b$ and the lower string a. After that the unit (bb, aa) from T_b has to be used exactly as many times as the unit (aa, a) has been applied before. Then an application of unit (bc, ab) is the sole possibility. This generates the upper strand $a^n b^n c$ and the lower strand $a^n b$. For the last part the units from T_c are used. Similarly as before, repeated applications of (cc, bb) yield the upper string $a^n b^n c^n$ and the lower string $a^n b^n$. So, it remains to complete the c's in the lower string. This is done by the units (c, bc), which can be applied only once, and (c, cc) which can be applied arbitrarily often. However, the derivation is successful only if the number of c's in the upper and lower strand match when the sole unit from E is applied. ∎

Clearly, the construction of Example 1 can be extended to an arbitrary number of symbols.

Example 2. Let Σ be an alphabet not containing the symbols $\{\$_1, \$_2, \$_3\}$. The SAS $S = \langle \Sigma \cup \{\$_1, \$_2, \$_3\}, A, T, E \rangle$ generates the non-context-free language $\{ \$_1 w \$_2 w \$_3 \mid w \in \Sigma^+ \}$, where for $x, y \in \Sigma$,

$$A = \{(\$_1, \$_1)\}, \quad T = T_1 \cup T_2 \cup T_3, \quad E = \{(\$_3, \$_3)\},$$
$$T_1 = \{(\$_1 x, \$_1), (xy, \$_1), (x\$_2, \$_1)\},$$
$$T_2 = \{(\$_2 x, \$_1 x), (xy, xy), (x\$_3, x\$_2)\},$$
$$T_3 = \{(\$_3, \$_2 x), (\$_3, xy), (\$_3, x\$_3)\}.$$

Similar as in Example 1, the units from T_1 are used to generate the upper strand $\$_1 w \$_2$ and the lower strand $\$_1$. Then units from T_2 are assembled to generate the upper strand $\$_1 w \$_2 w \$_3$ and the lower strand $\$_1 w \$_2$. Finally, we obtain $\$_1 w \$_2 w \$_3$ as upper and lower strand by the units in T_3. ∎

The construction of Example 2 can be extended to an arbitrary number of copies of the word w.

Sticker Systems. Now let us turn to sticker systems. In their original definitions also complete double strands are generated, where the projection on one of the strands is the string language generated. However, inspired by DNA the symbols at the same positions in the strands have to be in relation with respect to a given symmetric (complementarity) relation on the underlying alphabet. So, in general, the upper and lower strand of a generated double strand may be different. But it has been shown in [5] that every string language generated by a sticker system is generated by such a system with a one-to-one complementarity relation as well. Since here we are interested in the generative capacity, we safely may assume that the complementarity relation is always the identity to simplify the presentation.

Since general sticker systems generate double strands beginning with an axiom by assembling fragments at both ends of the currently derived double strand, we have to introduce some notation. Moreover, since in the literature of sticker systems the double strands are written one at the top of each other, we stick with this presentation, which also helps to distinguish between parts belonging to string assembling systems and parts belonging to sticker systems.

For sticker systems, we consider only (fragments) of double strands where either one or both strands are empty, or where there is a complete part with sticky ends at the left and at the right. So, we denote the set of all complete matching double strands over an alphabet Σ by $D_{\Sigma}^{=}$ and write $\begin{bmatrix} w \\ w \end{bmatrix}$ for the element of $D_{\Sigma}^{=}$ with strands $w \in \Sigma^+$. The sticky ends to the left and/or right of an element of $D_{\Sigma}^{=}$ are substrings that do not have a matching part in the opposite strand. They may appear at the upper or lower strand. For $w \in \Sigma^+$, we write $\begin{pmatrix} \lambda \\ w \end{pmatrix}$, $\begin{pmatrix} w \\ \lambda \end{pmatrix}$, and $\begin{pmatrix} \lambda \\ \lambda \end{pmatrix}$ for such incomplete double strands and denote their set by E_{Σ}. For easier writing, we sometimes simply write λ for $\begin{pmatrix} \lambda \\ \lambda \end{pmatrix}$. The set of complete or incomplete double strands with sticky ends to the left and/or right having or not having a complete part is denoted by $D_{\Sigma}^{\overline{\overline{=}}} = E_{\Sigma} \cup E_{\Sigma} \cdot D_{\Sigma}^{=} \cdot E_{\Sigma}$. The elements of $D_{\Sigma}^{\overline{\overline{=}}}$ are called *dominoes*.

A sticker system is a triple $S = \langle \Sigma, A, D \rangle$, where Σ is a finite, nonempty set of *symbols*, $A \subset E_\Sigma \cdot D_\Sigma^{\overline{\overline{}}} \cdot E_\Sigma$ is a finite set of *axioms*, and $D \subset D_\Sigma^{\overrightarrow{}} \times D_\Sigma^{\overleftarrow{}}$ is a finite set of *pairs of dominoes* over the alphabet Σ.

Let $u \in D_\Sigma^{\overline{\overline{}}}$ be the double strand $\binom{l_0}{l_1} \begin{bmatrix} v \\ v \end{bmatrix} \binom{r_0}{r_1}$ with $v \in \Sigma^+$ and at least one sticky end from $\{l_0, l_1\}$ and from $\{r_0, r_1\}$ empty, respectively, and $(d_0, d_1) \in D$ be a pair of dominoes. A derivation step $u \Rightarrow u'$ by using the domino pair (d_0, d_1) consists of assembling d_0 at the left and d_1 at the right of u. Basically, assembling means to concatenate the upper substring of the domino to the upper strand of u and the lower substring of the domino to the lower strand of u without moving the strands against each other. This assembling is possible only if the result of the concatenation is again a double stranded sequence from $D_\Sigma^{\overline{\overline{}}}$. Otherwise the domino is not applicable to u. Exemplarily, we present two possibilities of assembling dominoes formally. A complete treatment can be found in [11].

So, assembling $d_0 = \binom{l_0'}{l_1'} \begin{bmatrix} v' \\ v' \end{bmatrix} \binom{r_0'}{r_1'}$ with $v' \in \Sigma^+$ and at least one sticky end from $\{l_0', l_1'\}$ and one from $\{r_0', r_1'\}$ being empty, at the left of u results in $\binom{l_0'}{l_1'} \begin{bmatrix} v'r_0'l_0v \\ v'r_1'l_1v \end{bmatrix} \binom{r_0}{r_1}$, if and only if $r_0'l_0 = r_1'l_1$.

Assembling $d_1 = \binom{v'}{\lambda}$ at the right of u results in $\binom{l_0}{l_1} \begin{bmatrix} vv'' \\ vv'' \end{bmatrix} \binom{r_0'}{r_1'}$, if and only if $v'' = \lambda$, $r_0' = r_0v'$, and $r_1' = \lambda$ if $r_1 = \lambda$, and $v'' = r_1$, $r_1' = \lambda$ if $v' = r_1r_0'$ and $r_0 = \lambda$, and $v'' = v'$, $r_0' = \lambda$ if $r_1 = v'r_1'$ and $r_0 = \lambda$.

The language generated by S is defined to be

$$L(S) = \left\{ w \in \Sigma^+ \mid u \Rightarrow^* \begin{bmatrix} w \\ w \end{bmatrix}, \text{ for some } u \in A \right\}.$$

Several important types of restricted sticker systems have been considered. The maximal overhang of a double strand $u \in D_\Sigma^{\overline{\overline{}}}$ is called the *delay* of u and it is denoted by $d(u)$. A derivation $u_1 \Rightarrow u_2 \Rightarrow \cdots \Rightarrow u_k$ with $u_1 \in A$ and $u_k \in D_\Sigma^{\overline{\overline{}}}$ is said to be *of delay* m, if the maximal length of overhang for every u_i $1 \leq i \leq k$ is at most m.

A sticker system $S = (\Sigma, A, D)$ is *one-sided* if for each pair $(d_0, d_1) \in D$ either $d_0 = \lambda$ or $d_1 = \lambda$, *regular* if for each pair $(d_0, d_1) \in D$ we have $d_0 = \lambda$, and *simple* if for each pair $(d_0, d_1) \in D$ we have $d_0, d_1 \in \binom{\Sigma^*}{\lambda} \cup \binom{\lambda}{\Sigma^*}$.

Following the notation in [11], we denote by ASL the family of languages generated by general sticker systems, by OSL, RSL, and SSL the families of languages generated by one-sided, regular, and simple sticker systems, respectively, and use the notation SOSL and SRSL for the languages families generated by simple one-sided and simple regular systems. For all these families we distinguish between unrestricted derivations and derivations with bounded delay. So, we add (n) or (b) to the notation to indicate non-restricted and bounded-delay derivations (see Fig. 2 for the hierarchy of generated language families). Since being *bounded delay* is a property of derivations and not a structural property of the sticker system, we have to clarify what bounded-delay means. We say that a sticker system is of bounded delay, if for each word generated by the system there exists a (possibly different) derivation of bounded delay.

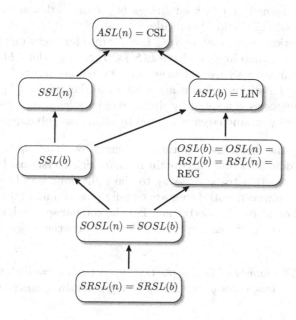

Fig. 2. Hierarchy of language families generated by sticker systems. A single arrow means strict inclusion. All families not connected by a path are incomparable.

3   Comparison of SAS with Sticker Systems

In this section, we compare the generative capacities of string assembling systems with those of different types of sticker systems.

The weakest variant of sticker systems are simple regular sticker systems. They generate a subregular family of languages. For example, the regular language $\{ a^n b \mid n \geq 1 \}$ cannot be generated by any of these systems.

Theorem 3. *The family of languages* $SRSL(b) = SRSL(n)$ *is a proper subfamily of the family of languages generated by string assembling systems.*

Proof. Let $S = \langle \Sigma, A, D \rangle$ be a simple regular sticker system. We construct a string assembling system $S' = \langle \Sigma, A', T', E' \rangle$ such that $L(S) = L(S')$. There are two main restrictions of regular simple sticker systems. First, dominoes can be appended only on the right-hand side and, second, either the upper or the lower strand of each domino is empty. Thus, we define for each axiom from A of the form $\begin{pmatrix} \lambda \\ \lambda \end{pmatrix} \begin{bmatrix} v \\ v \end{bmatrix} \begin{pmatrix} r_0 \\ r_1 \end{pmatrix}$ an axiom $(vr_0, vr_1) \in A'$. For every pair of dominoes

$(d_0, d_1) \in D$, $d_0 = \begin{pmatrix} \lambda \\ \lambda \end{pmatrix}$ and $d_1 = \begin{pmatrix} r_0 \\ r_1 \end{pmatrix}$ with either $r_0 = \lambda$ or $r_1 = \lambda$, we define units $(xr_0, yr_1) \in T'$, for every symbols $x, y \in \Sigma$. The set of ending units is defined by $(x, x) \in E'$, for every $x \in \Sigma$. Since the derivation processes of S and S' are both defined to be one-way to the right and, further, for every domino in D, units are defined in T' whose strings begin with arbitrary symbols, the derivation processes of S and S' coincide.

The construction shows that $L(S) \subseteq L(S')$. Since for every unit constructed there is a unique domino in S, we have $L(S') \subseteq L(S)$ and, thus, $L(S) = L(S')$.

On the other hand, SAS are able to generate even non-context-free languages, while simple regular sticker systems are a subset of the regular languages. Thus the family of languages generated by simple regular sticker systems is a proper subset of the family of languages generated by string assembling systems.   □

Climbing up the hierarchy of sticker languages by giving up the restriction to have simple dominoes only, we obtain regular sticker systems. Relaxing furthermore the restriction to be one-way to the right only, we obtain one-sided sticker systems. However, both types are equally powerful and for every derivation there is a bounded-delay derivation. For the comparison with string assembling systems this step of relaxing crosses the edge between strict inclusion and incomparability.

Theorem 4. *The family of languages $OSL(b) = OSL(n) = RSL(b) = RSL(n)$ and the family of languages generated by string assembling systems are incomparable.*

Proof. Since $OSL(n)$ coincides with the family of regular languages [11], one-sided sticker systems can generate the language $\{a\} \cup \{a^{2n} \mid n \geq 2\}$, which cannot be generated by any string assembling system [7].

Conversely, the language $\{\$_1 w \$_2 w \$_3 \mid w \in \{a, b\}^+\}$ of Example 2 shows that there are even non-context-free languages generated by string assembling systems. Since these cannot belong to $OSL(n)$, the theorem follows.   □

The following example is used to show that there is a language generated by SAS that cannot be generated by any simple sticker system.

Example 5. The language $aL^+ \cup aL^+a^*\$ \cup a^+\$ \cup \{a\}$, where L is defined as $L = \{a^{n-1}b^n c^m a^m \mid m, n \geq 1\}$, is generated by the string assembling system $S = \langle \Sigma, A, T, E \rangle$ with single axiom (a, a), set of ending units $E = \{(a, a), (\$, \$)\}$, and assembling units as follows.

1. $(aa, a) \in T$	6. $(bb, aa) \in T$	11. $(c, ca) \in T$
2. $(a\$, a) \in T$	7. $(bc, ab) \in T$	12. $(cc, aa) \in T$
3. $(\$, aa) \in T$	8. $(c, bb) \in T$	13. $(ca, a) \in T$
4. $(\$, a\$) \in T$	9. $(c, bc) \in T$	
5. $(ab, a) \in T$	10. $(c, cc) \in T$	

The singleton $\{a\}$ is generated by the axiom and the ending unit (a, a). The units (1)–(4) are used to generate words of the form $\{a^n\$ \mid n \geq 1\}$. To this end, first unit (1) is iteratively applied which yields (a^n, a). Then, by applying unit (2) one gets $(a^n\$, a)$, and the completion of the lower strand is initialized, which is done by units (3) and (4). Clearly, it is not possible to apply any other unit after unit (2).

The derivation of words of the form aL^+ also starts with applications of unit (1). Now with unit (5) one obtains $(a^n b, a)$, and by units (6)–(9) we get $(a^n b^n c, a^n b^n c)$. Next, unit (10) prolongs the lower strand by c^m, and units (11)–(13) are used to derive $(a^n b^n c^m a, a^n b^n c^m a^m)$. After completing the upper strand by unit (1), we have $(a^n b^n c^m a^m, a^n b^n c^m a^m)$. Now the derivation process can either stop with the ending unit (a, a) or can be repeated.   ∎

Theorem 6. *The families of languages SSL(b) and SSL(n) are incomparable with the family of languages generated by string assembling systems.*

Proof. The dominoes of simple sticker systems have either an empty upper strand or an empty lower strand. We consider the language $L' = aL^+ \cup aL^+ a^* \$ \cup a^+ \$ \cup \{a\}$ with $L = \{a^{n-1} b^n c^m a^m \mid m, n \geq 1\}$ of Example 5, which is generated by a string assembling system. In contrast to the assertion, assume that L' is generated by a simple sticker system $S = \langle \Sigma, A, D \rangle$.

An immediate observation for simple sticker systems is as follows. If there is some derivation from a double strand u to v applying the pairs of dominoes $p_1, p_2, \ldots, p_\ell$, where all dominoes contain unary strands over the same symbol, then applying the same pairs of dominoes in any other order yields a derivation from u to v as well.

First, we consider words from $L'' = \{a^n \$ \mid n \geq 1\} \subset L'$. In every derivation of a word $w \in L''$ at most two times a domino is assembled whose strands are not of the form a^*. Since the number of axioms in A and the number of dominoes in D is finite, there are at most two fixed pairs of dominoes q_1 and q_2 whose strands contain a $\$$, and a fixed axiom d_0 such that infinitely many words from L'' are generated by a derivation that applies d_0, q_1, q_2, and further pairs of dominoes whose strands are of the form a^*. We denote the set of these words by $L''_{d_0, q_1, q_2} \subseteq L''$.

Let $\{p_1, p_2, \ldots, p_m\}$ be the set of all pairs of dominoes from D whose strands are of the form a^*. For any word $w \in L''_{d_0, q_1, q_2}$, we choose a derivation and consider the multiplicities of applications of the pairs of dominoes p_i. So, let $s_w = (i_1, i_2, \ldots, i_m)$, with $0 \leq i_j$, for $1 \leq j \leq m$, denote these multiplicities (for the chosen derivation). Since L''_{d_0, q_1, q_2} is infinite, there are at least two different words $w, w' \in L''_{d_0, q_1, q_2}$ such that $s_w = (i_1, i_2, \ldots, i_m)$ and $s_{w'} = (i_1 + \ell_1, i_2 + \ell_2, \ldots, i_m + \ell_m)$, for $0 \leq \ell_j, 1 \leq j \leq m$. Since d_0, q_1, and q_2 are fixed, we conclude that assembling ℓ_1 times p_1, ℓ_2 times p_2, ..., and ℓ_m times p_m adds complete double strands with single strands of the form a^* to the left and to the right of the axiom. Moreover, at least one of these double strands is non-empty.

Now we come back to the language L' and choose some word $z = a^n b^n c^m a^m \in aL \subset L'$, with m, n long enough. Then the derivation of z can be extended by assembling ℓ_1 times p_1, ℓ_2 times p_2, $\ldots$, and ℓ_m times p_m. The result is again a complete double strand and, thus, the generated word belongs to L'. But this is a contradiction to the definition of L', since it does not contain any word $a^{n+k_1}b^n c^m a^{m+k_2}$ with $k_1 + k_2 \geq 1$. Therefore, L' cannot be generated by any simple sticker system.

On the other hand, the language $\{\, w\$w^R \mid w \in \{a, b\}^* \,\}$ is obviously generated by a simple sticker system, but cannot be generated by any SAS [7].   □

Next, we relax the restriction of simple dominoes and turn to general sticker systems. It has been shown in [11] that the family of languages ASL(b) generated by general sticker systems with derivations of bounded delay coincides with the family of linear context-free languages (LIN).

Theorem 7. *The family of languages ASL(b) is incomparable with the family of languages generated by string assembling systems.*

Proof. We have ASL(b) = LIN $\supset$ REG. Since the family of languages generated by string assembling systems and REG are incomparable, there is a language generated by a general sticker system with bounded delay, that cannot be generated by any SAS (for example the language $\{a\} \cup \{\, a^{2n} \mid n \geq 2 \,\}$ [7]). On the other hand, the language $L = \{\, a^n b^n c^n \mid n \geq 1 \,\}$ of Example 1 shows that there are non-context-free languages generated by SAS. So, the incomparability follows.   □

The results above show that even if the concepts of sticker systems and string assembling systems are similar, their computational power differs essentially. But the next representation theorem shows that there are also deep connections between both models.

Theorem 8. *For any SAS S over an alphabet Σ, a sticker system $\hat{S}$ can effectively be constructed over an alphabet $\hat{\Sigma}$ such that there is an injective mapping $f \colon \Sigma^* \to \hat{\Sigma}^*$ and $L(\hat{S}) = \{f(w)\#w \mid w \in L(S)\}$.*

4   Decidability

In this section we consider decidability questions for the family of language generated by string assembling systems. In [7] it has already been shown that the questions of emptiness, finiteness, infiniteness, equivalence, inclusion, regularity, and context-freeness are undecidable. The proof is by reduction of Post's Correspondence Problem (PCP). It is known that the PCP is still undecidable, if the length of the words is limited to two [2]. Thus, the undecidability proof in [7] requires the construction of SAS whose units are at least 3-length-restricted. On the other hand, string assembling systems that are 1-length-restricted are not productive and generate finite languages only. Clearly, in this case the mentioned problems are decidable. The decidability status of 2-length-restricted SAS remained open, since for the PCP with words of length 1, emptiness is decidable. Here we solve the open status by showing that even for 2-length-restricted

SAS all the aforementioned problems are undecidable. Furthermore, we can show that they are not even semi-decidable. To this end, we turn to reductions from one-sided Turing machine computations. The histories of Turing machine computations are encoded into strings, a technique introduced in [3]. Here we consider deterministic Turing machines with one single one-sided tape and one single read-write head.

Let $M = \langle Q, \Sigma, T, q_0, \sqcup, \rhd, F, \delta \rangle$ be an one-sided Turing machine where Q is the finite set of states, T is the finite set of tape symbols, $\Sigma \subset T$ is the set of input symbols, q_0 is the initial state, $\sqcup \in T \setminus \Sigma$ is the blank symbol, $\rhd \in T \setminus \Sigma$ is the left endmarker, $F \subseteq Q$ is the finite set of accepting states and $\delta \colon Q \times T \to Q \times (T \cup \{L, R\})$ is the transition function.

A configuration of M can be written as a string of the form $\rhd T^* Q T^*$ such that $\rhd x_1 x_2 \cdots x_i q x_{i+1} \cdots x_n$ is used to express that M is in state q, scanning tape symbol x_i, and $x_1 x_2 \cdots x_n$ is the (non-blank) tape inscription.

Without loss of generality and for technical reasons, we assume that the Turing machines cannot print blanks, may not leave blank fields, make at least one move, halt when entering an accepting state, and move their head to the right on reading the endmarker. Whenever, the head is moved to the right of the last symbol, a blank is appended to the tape inscription (that must be rewritten in the next step).

For the construction of an SAS, we need an additional technical transformation of the representation of configurations. We combine the current state q and the symbol x on its left to some metasymbol $[x, q]$ and mark the symbols on the right of the metasymbol. Thus, let f be a mapping such that $f(\rhd x_1 x_2 \cdots x_i q x_{i+1} \cdots x_n) = \rhd x_1 x_2 \cdots x_{i-1} [x_i, q] x'_{i+1} \cdots x'_n$, where $q \in Q$, $n \geq i \geq 1$, and $x_j \in T$, for $1 \leq j \leq n$. The configuration representing the empty word is encoded as $f(\rhd q) = [\rhd, q]$.

Dependent on M we now define the language of valid computations. Let $\#_1, \#_2, \$ \notin T \cup Q$ and $w_i \in \rhd T^* Q T^*$, $0 \leq i \leq m$, be configurations of M. Then VALC(M) is defined to be the language of all words of the form

$$\#_1 f(w_0) \$ f(w_1) \$ \cdots \$ f(w_m) \$ \#_2,$$

where w_0 is an initial configuration of the form $\rhd \Sigma^* q_0$, w_m is an accepting and, thus, halting configuration, and w_i is the successor configuration of w_{i-1}, $1 \leq i \leq m$.

Theorem 9. *Emptiness is not semidecidable for 2-length-restricted SAS.*

Proof. Given a one-sided Turing machine $M = \langle Q, \Sigma, T, q_0, \sqcup, \rhd, F, \delta \rangle$ we construct a 2-length-restricted SAS $\hat{S} = \langle \hat{\Sigma}, \hat{A}, \hat{T}, \hat{E} \rangle$ generating VALC(M) as follows. First, we define $T_p = \{ x' \mid x \in T \}$, $T_s = \{ [a, q] \mid a \in T, q \in Q \}$, and set $\hat{\Sigma} = T \cup T_p \cup T_s \cup \{\#_1, \#_2, \$\}$. The sole axiom is defined by the unit $(\#_1, \#_1)$. We continue with units that generate the upper strand of the first configuration. For all $a \in \Sigma$ and $x \in \Sigma \cup \{\rhd\}$,

1. $(\#_1 \triangleright, \#_1) \in \hat{T}$
2. $(x[a, q_0], \#_1) \in \hat{T}$
3. $(xa, \#_1) \in \hat{T}$
4. $(\#_1[\triangleright, q_0], \#_1) \in \hat{T}$

5. $([x, q_0]\$, \#_1) \in \hat{T}$
6. $(\$\triangleright, \#_1\triangleright) \in \hat{T}$
7. $(\$\triangleright, \#_1[\triangleright, q_0]) \in \hat{T}$ if $\delta(q_0, \triangleright) = (q', R)$.

The units (1)–(3) can be used to continue after the derivation of the axiom. Unit (4) covers the special case where the input word is empty. If unit (1) is applied, the upper strand can be extended further by unit (3) to a double strand of the form $(\#_1 w, \#_1), w \in \triangleright \Sigma^*$. At some point the last symbol together with the state is derived by unit (2). Unit (5) completes the upper strand of the first block $(\#_1 w[a, q_0]\$, \#_1)$, while unit (6) and (7) realize the transition to the next block.

From now on let $x, y, a, b \in T$ be symbols, $x', y', a', b' \in T_p$ be marked symbols, and $q, q' \in Q$ be states. The basic units that generate stepwise the configurations of M are defined by

8. $(xa, xa) \in \hat{T}$ 9. $(x'a', x'a') \in \hat{T}$ 10. $(a'\$, a'\$) \in \hat{T}$.

All units (8)–(10) are used to copy the parts of the configuration that stay unchanged.

Let $\delta(q, a) = (q', b)$. Then the write operation of M is generated by

11. $(x[b, q'], x[a, q]) \in \hat{T}$
12. $([b, q']\$, [a, q]\$) \in \hat{T}$

13. $([b, q']x', [a, q]x') \in T'$
14. $(\$[b, q'], \$[a, q]) \in \hat{T}$.

Let $\delta(q, a) = (q', R)$. Then the move right operation without a new blank symbol is generated by

15. $(xa, x[a, q]) \in \hat{T}$
16. $(a[x, q'], [a, q]x') \in \hat{T}$

17. $([x, q']a', x'a') \in T'$
18. $([x, q']\$, x'\$) \in \hat{T}$.

The upper strand is extended by a normal symbol, while the lower strand is extended by a metasymbol symbol with unit (15). The next appropriate derivation step is unique, since now the last symbol of the upper strand is a normal symbol while the last symbol in the lower strand is a metasymbol and unit (16) is the single unit that can be applied. Afterwards, unit (17) is the sole possibility. Unit (18) treats the special case when the next symbol is the last symbol of the configuration.

The extension by a blank symbol is a further special case. In principle the idea is the same as above. The only difference is that the blank symbol has to be added in the metasymbol (unit (16)) and the need of the $ symbol afterwards by units (19) and (20). Let $\delta(q, a) = (q', R)$. Then

19. $(a[\sqcup, q'], [a, q]\$) \in \hat{T}$ 20. $([\sqcup, q']\$, \$) \in \hat{T}$.

Furthermore we consider the move left operations. Let $\delta(q, y) = (q', L)$. Then

21. $(x[a, q'], xa) \in \hat{T}$ 23. $(y'x', [y, q]x') \in \hat{T}$
22. $([a, q']y', a[y, q]) \in \hat{T}$ 24. $(y'\$, [y, q]\$) \in \hat{T}$.

Here the possibility of a nondeterministic choice is used. For the left move first the state symbol is added to the upper strand by unit (21), while in the lower strand a normal symbol is added. Then unit (22) adds a marked symbol in the upper strand and the metasymbol in the lower strand. By units (23) and (24) the connection to the rest of the marked strand is done.

The derivation of the last configuration is done by adding the special symbol $\#_2$ at the end. This enables the completion of the lower strand. Note that the construction is for accepting states only. Let $q_f \in F$. Then

25. $(\$\#_2, \$a) \in \hat{T}$ 28. $(\#_2, x[a, q_f]) \in \hat{T}$ 31. $(\#_2, a'\$) \in \hat{T}$,
26. $(\$\#_2, \$[a, q_f]) \in \hat{T}$ 29. $(\#_2, [a, q_f]x') \in \hat{T}$ 32. $(\#_2, [a, q_f]\$) \in \hat{T}$
27. $(\#_2, xa) \in \hat{T}$ 30. $(\#_2, a'x') \in \hat{T}$ 33. $(\#_2, \$\#_2) \in \hat{T}$.

The completion up to the state symbol is done by the units (25)–(28). Afterwards all symbols are marked by the units (29) and (30). The auxiliary symbols $\$\#_2$ are derived by the units (31)–(33). The sole ending unit is defined to be $(\#_2, \#_2) \in E$.

The construction shows that we can effectively construct a 2-length-restricted SAS $\hat{S}$ generating $\mathrm{VALC}(M)$ of an arbitrary given one-sided Turing machine M. Clearly, $L(M)$ is empty if and only if $L(\hat{S}) = \mathrm{VALC}(M)$ is empty. Since emptiness is not semi-decidable for deterministic one-sided Turing machines, the theorem follows.   □

By standard techniques one can easily show that the problems finiteness, infiniteness, equivalence, inclusion, regularity, and context-freeness are not semidecidable either.

Theorem 10. *Finiteness, infiniteness, equivalence, inclusion, regularity, and context-freeness are not semidecidable for 2-length-restricted SAS.*

5   Conclusion

In this paper, we considered string assembling systems and compared their generative capacity with those of several variants of sticker systems. Although the basic mechanisms of both types of systems seem to be closely related, their generative capacities differ essentially. While the copy language $\{\, \$_1 w \$_2 w \$_3 \mid w \in \Sigma^+ \,\}$ can be generated by some SAS, it has been shown that it is not generated by the most variants of sticker systems. Conversely, many variants of the mirror language $\{\, w \mid w \in \{a, b\}^* \text{ and } w = w^R \,\}$ are generated by variants of sticker systems (that can generate all linear context-free languages), but cannot be generated by any SAS, since it cannot be accepted by any nondeterministic one-way two-head finite automaton. So, sticker systems can handle mirrored inputs while SAS cannot. It turned out that string assembling systems are incomparable with most variants of sticker systems. An exception is the weakest type of sticker system that may only assemble simple dominoes on the right-hand side. Its generative capacity is strictly weaker than the generative capacity of string assembling systems. However, there are two problems left open. It is known that the strongest variant of sticker systems can generate all linear context-free languages and, thus, can generate a language that cannot be generated by string assembling systems. But the converse is open. Candidate languages are the copy language $\{\,\$_1w\$_2w\$_3 \mid w \in \Sigma^+\,\}$ and $\{\,a\$a^2\$a^4\$\dots\$a^{2^i}\# \mid i > 1\,\}$. We conjecture that they are witnessing the incomparability also in this case. The second open problem is the relation between string assembling systems and sticker systems that are one-sided but still restricted to simple dominoes.

Finally, we considered decidability questions for the family of language generated by string assembling systems. It was known before that the problems of emptiness, finiteness, infiniteness, equivalence, inclusion, regularity, and context-freeness are undecidable if the units of the string assembling systems have at least length 3. Since 1-length-restricted string assembling systems are not productive and generate finite languages only, the 2-length-restricted was open. Here we solve the open status by showing that even for 2-length-restricted SAS all the aforementioned problems are undecidable. Furthermore, we can show that they are not even semi-decidable which improves the previously known results furthermore.

References

1. Freund, R., Păun, G., Rozenberg, G., Salomaa, A.: Bidirectional sticker systems. In: Pacific Symposium on Biocomputing (PSB 1998), pp. 535–546. World Scientific, Singapore (1998)
2. Halava, V., Hirvensalo, M., de Wolf, R.: Marked PCP is decidable. Theor. Comput. Sci. **255**, 193–204 (2001)
3. Hartmanis, J.: Context-free languages and Turing machine computations. In: Proceedings of the Symposia in Applied Mathematics, vol. 19, pp. 42–51 (1967)
4. Kari, L., Păun, G., Rozenberg, G., Salomaa, A., Yu, S.: DNA computing, sticker systems, and universality. Acta Inf. **35**, 401–420 (1998)
5. Kuske, D., Weigel, P.: The role of the complementarity relation in Watson-Crick automata and sticker systems. In: Calude, C.S., Calude, E., Dinneen, M.J. (eds.) DLT 2004. LNCS, vol. 3340, pp. 272–283. Springer, Heidelberg (2004). https://doi.org/10.1007/978-3-540-30550-7_23
6. Kutrib, M., Wendlandt, M.: Bidirectional string assembling systems. In: Non-Classical Models of Automata and Applications (NCMA 2012). books@ocg.at, vol. 290, pp. 107–121. Austrian Computer Society (2012)
7. Kutrib, M., Wendlandt, M.: String assembling systems. RAIRO Inf. Théor. **46**, 593–613 (2012)

8. Kutrib, M., Wendlandt, M.: Parametrizing string assembling systems. In: Câmpeanu, C. (ed.) CIAA 2018. LNCS, vol. 10977, pp. 236–247. Springer, Cham (2018). https://doi.org/10.1007/978-3-319-94812-6_20

9. McNaughton, R.: Algebraic decision procedures for local testability. Math. Syst. Theor. **8**, 60–76 (1974)

10. Păun, G., Rozenberg, G.: Sticker systems. Theor. Comput. Sci. **204**, 183–203 (1998)

11. Păun, G., Rozenberg, G., Salomaa, A.: DNA computing: new computing paradigms. In: Texts in Theoretical Computer Science, Springer, Berlin (1998). https://doi.org/10.1007/978-3-662-03563-4

12. Post, E.L.: A variant of a recursively unsolvable problem. Bull. AMS **52**, 264–268 (1946)

13. Zalcstein, Y.: Locally testable languages. J. Comput. Syst. Sci. **6**, 151–167 (1972)

Fractal Dimension of Assemblies in the Abstract Tile Assembly Model

Daniel Hader[1], Matthew J. Patitz[1] , and Scott M. Summers[2]([✉])

[1] University of Arkansas, Fayetteville, AR 72703, USA
{dhader,patitz}@uark.edu
[2] University of Wisconsin-Oshkosh, Oshkosh, WI 54901, USA
summerss@uwosh.edu

Abstract. In this paper, we investigate the power of systems in the abstract Tile Assembly Model to self-assemble shapes having fractal dimensions between 1 and 2. We introduce a concept of sparsity as a tool for investigating such systems and demonstrate its utility by proving how it relates to fractal dimension.

1 Introduction

Algorithmic, tile-based self-assembly has been a very successful topic of research with ramifications in nanoscience, medicine, computer science, etc. Ideas from this field have been used to construct precise nanoscale structures utilizing the dynamics of DNA base pairing to emulate devices such as binary counters and logic gates [7,27]. Analysis of models such as the abstract Tile Assembly Model (aTAM) and 2-Handed Assembly Model (2HAM) have demonstrated that tile-based self-assembly is capable of universal computation and arbitrary shape construction, among many other complex algorithmic tasks [2,4,6,24–26]. Additionally, these models have been shown to induce rich hierarchies similar to those studied in computational complexity theory. Investigating the limits and capabilities of these models therefore is an important academic effort both theoretically and practically. Because these models are inherently geometrical, requiring that tiles occupy explicit locations in space, questions regarding the kinds of shapes that can be constructed from tiles are of central importance. Fractals are a particularly interesting class of shapes because of their inherently mathematical definitions and recursive nature. Additionally, whereas familiar shapes such as lines and squares occupy an integer value of spacial dimensions, fractals can be assigned a non-integer fractal dimension which informally measures how the area scales when considering larger and larger portions of the fractal. Consequently, fractals, particularly a class of fractals called discrete self-similar fractals (DSSFs), have been the topic of focus for numerous investigations in the algorithmic tile-assembly literature.

When considering weak self-assembly, where some subset of tiles are allowed to occupy the negative space of the desired shape, it has been demonstrated

M. J. Patitz—This work was supported in part by National Science Foundation grant CAREER-1553166.

© Springer Nature Switzerland AG 2021
I. Kostitsyna and P. Orponen (Eds.): UCNC 2021, LNCS 12984, pp. 116–130, 2021.
https://doi.org/10.1007/978-3-030-87993-8_8

that large classes of fractal shapes can be constructed theoretically [14,15] and even physically [23]. Strict self-assembly of fractals on the other hand, where tiles can only occupy space belonging to the desired shape, is a more difficult problem. In signal passing models wherein the attachment of tiles can affect the structure of others nearby, it has been shown that all DSSFs, those fractals that can be defined recursively bottom-up from a finite generator shape, can be constructed [11,19]. In hierarchical models, where subassemblies of tiles can join to form larger assemblies, it has been shown that some classes of fractals can be assembled [3,10,12] while some cannot [2]. In the aTAM, the original model of tile-assembly [26] wherein tiles attach one by one to a seed assembly, it has been shown that many classes of fractals cannot be self-assembled [1,8,10,17].

The question of whether or not any discrete self-similar fractal (with non-trivial dimension strictly between 1 and 2) can strictly self-assemble in the aTAM has long been open. It has been shown that the Sierpinski triangle cannot strictly self-assemble in the aTAM [17], the proof of which essentially boils down to a counting argument. Additionally, it has been shown that weakening DSSFs to their fibered [17,20] or laced versions [15,18], wherein spacing between stages grows logarithmically rather than remaining constant or counters are embedded in the holes of fractal stages, allows for strict self-assembly of large classes of such fractals. (Interestingly, these fibered and laced versions have the same fractal dimension as their originals.) It is generally conjectured that DSSFs cannot strictly self-assemble in the aTAM. If this is the case, then informally, these results seem to imply that the reason is because DSSFs are too sparse to allow tiles to pass around the information necessary to distinguish between fractal stages and has little to do with the fractal dimension of the shape itself. To investigate this idea, in this paper we formally introduce definitions of density and sparsity which largely mirror the concept of *natural density* of sequences of natural numbers. We then prove some results regarding the relationship between sparsity and ζ-dimension, a notion of fractal dimension which is naturally applicable to discrete shapes (see [5] for a thorough discussion regarding ζ-dimension). Furthermore, we show that sparsity acts as a natural distinguishing feature between DSSFs and their fibered/laced counterparts, whereas ζ-dimension does not, suggesting that our definitions might be useful tools for investigating impossibility of strict self-assembly of DSSFs in the aTAM. Finally, to provide further evidence, though indirectly, that it is sparsity and not ζ-dimension which makes shapes impossible to strictly self-assemble in the aTAM, we construct a universal aTAM tileset, which is able to construct non-sparse shapes with any desired ζ-dimension from a large class of possible values that includes all algebraic numbers.

2   Preliminaries

Here we provide some preliminary definitions and concepts that will be useful for understanding our results. This paper is particularly concerned with shapes that self-assemble in the abstract Tile Assembly Model (aTAM). Here we give a brief informal overview of the aTAM and provide necessary definitions. For a more formal treatment, see [17,24,26].

2.1   The Abstract Tile Assembly Model

A *tile type* is a unit square whose 4 sides may each contain a glue. A *glue* consists of a *label* (which is often represented by a finite string or color) and a positive integer value called its *strength*. A *tile* is an instance of a tile type which can be thought of as occupying a location in the integer lattice $\mathbb{Z}^2$. Two adjacent tiles are said to be *bound* or *attached* with strength s if their abutting glues share the same label l and strength s. In this case, the abutting glues are said to be *matching*. Informally an *assembly* is a collection of bound tiles and we say that an assembly is τ-*stable* if each tile or subassembly is attached with at least strength τ to the rest of the assembly. More formally we define an assembly to be a partial function $\alpha : \mathbb{Z}^2 \to T$ where T is a finite set of tile types. The *binding graph* of an assembly α is the subgraph of the integer lattice graph whose vertices correspond to the domain of α and edges represent bound glues between tiles. An assembly is τ-stable if for every cut of its binding graph, the edges cut correspond to glues whose strengths sum to at least τ.

Definition 1. *A* tile assembly system (TAS) *is a triple* $\mathcal{T} = (T, \sigma, \tau)$ *where*

- T *is a finite set of distinct tile types called the* tileset,
- σ *is a finite, τ-stable assembly called the* seed assembly, *and*
- τ *is a positive integer called the* binding threshold.

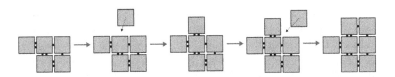

Fig. 1. An illustration of tiles attaching to an assembly. The first tile attachment happens using a single strength 2 glue whereas the second takes advantage of cooperation, using 2 strength 1 glues. Here it's assumed that the binding threshold is 2.

aTAM Dynamics. Given a TAS $\mathcal{T} = (T, \sigma, \tau)$, we say that assembly α' is *producible* from assembly α if α' is identical to α with the addition of a single tile from T and both are τ-stable. That is, α' results from α by the attachment of a single tile. The binding threshold τ limits the attachments to those whose strength meets or exceeds τ. If for example $\tau = 1$ then any pair of matching glues would be sufficient for attachment of a tile to an assembly. If $\tau > 1$ however, then certain attachments might not be possible. A single strength 1 glue will not allow a tile to attach in this case, but a single glue of strength τ will. Additionally, if multiple glues on a single tile match with the assembly and their sum is at least τ, then the tile can attach and this attachment is called *cooperative*. Cooperative binding is sufficient and necessary for complex Turing complete behavior in the aTAM (see Fig. 1 for an example).

An *assembly sequence* is a sequence of assemblies $\vec{\alpha} = \{\alpha_0, \alpha_1, \ldots\}$, which may or may not be finite, such that each assembly is producible from the previous one. An assembly is called *terminal* if no other assemblies are producible from it. An assembly α is a *terminal assembly of* T if there exists an assembly sequence beginning with σ and ending with α, or in the case of an infinite assembly sequence limiting to α.

Useful aTAM Gadgets. The definitions presented above informally describe the aTAM and its dynamics. There are many more definitions and notational conventions that are useful and often necessary when discussing the aTAM; however, in this paper we will focus little on the details of aTAM constructions and instead take advantage of commonly used gadgets in the aTAM. These gadgets have been developed previously in the literature and our constructions will be described in terms of these gadgets in much the same way that defining the behavior of a Turing machine is often done with pseudo-code rather than defining a large, complex transition function. For more information regarding common aTAM gadgets and constructions, refer to the aTAM article on self-assembly.net. The two gadgets we will make particular use of are *Turing machine gadgets* and *planters*.

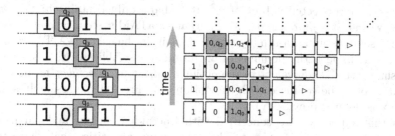

Fig. 2. An example Turing machine (left) and the corresponding Turing machine gadget (right) implemented in the aTAM with binding threshold 2. Each row represents the configuration of the tape at consecutive time steps. Notice that the tiles corresponding to the head attach in the corresponding locations along the tape in each row. Special tiles along the far end of the tape cause it to grow each row so that there can be an effectively infinite tape.

Turing Machine Gadgets. A Turing machine gadget simulates some fixed Turing machine whose definition is encoded in the tileset. Each row of the Turing machine gadget corresponds to the tape at some time step. Since Turing machine tapes are infinite, we define these gadgets so that the number of tiles in each row grows when necessary to accommodate any number of tape cells the Turing machine might need. The state and head position is passed along using special glues and the transition function is embedded in the definitions of the glues. For our purposes, we will require that the tiles that make up our Turing machine gadget form a square and thus we require special tiles to grow and fill in the remaining space after the Turing machine halts. While this is not typically desired behavior of such gadgets, it is not difficult to implement. Illustrated in Fig. 2 is an example Turing machine gadget. Detailed examples of such gadgets can be found in [16,21], among many others.

Planter Gadgets. Planters are gadgets which do little more than count in binary, starting from some initial value. Each column of the counter encodes a binary number in the glues of its tiles such that consecutive columns correspond to consecutive numbers. The number of tiles in a column of the counter is equal to the number of binary digits necessary to encode the specific number. Planters are simply counters which, at certain values (often at powers of 2, when the counter must expand by one bit), perform a rotation wherein a value is presented using some glue encoding orthogonal to the direction of the counter's increase. This rotation can be done using tiles with more complex glues in such a way that it occurs alongside the incrementing of the counter and requires no additional space. Additionally, a planter is capable of keeping track of more than just its counter value by having glues represent n-tuples of bits rather than just single bits. These values can also be rotated. Detailed examples of such gadgets can be found in [13,16], among many others.

2.2   Shapes and Fractals

A *shape* is a connected subset of $\mathbb{Z}^2$. If a shape only contains finitely many points, we call it a *finite* shape. We say that an aTAM system *produces* a shape if it has some terminal assembly whose domain is that shape. A TAS *uniquely produces* a shape if the domains of all terminal assemblies of that system are that shape (this is equivalent to the notion of *strict self-assembly*, formalized in [17] and not to be confused with the notion of a directed tile assembly system, which was defined previously).

   In this paper, we are concerned with fractal-like shapes and particularly the fractal dimension of shapes. There are many quantities that characterize fractal dimension and in well behaved instances these quantities often agree with each other; however, some are better suited for describing the fractal nature of certain classes of shapes than others. Because it has been traditionally used when discussing fractal shapes produced by the aTAM, we use the ζ-dimension as our notion of fractal dimension. Here we provide a basic definition that will be useful for our purposes. For a more in-depth overview of ζ-dimension, equivalent definitions, and related concepts please see [5].

ζ-Dimension. To facilitate with the definition of ζ-dimension and later in our definitions regarding sparsity, we define the *ball of radius r centered at $\vec{p}$* as the set of all points in $\mathbb{Z}^2$ with Euclidean distance from $\vec{p}$ no bigger than r. That is

Definition 2. *For any $r \in \mathbb{N}$ and $\vec{p} \in \mathbb{Z}^2$, the radius r ball centered at $\vec{p}$ is*

$$B_r\left(\vec{p}\right) =_{def} \{\vec{x} \in \mathbb{Z}^2 : \|\vec{x} - \vec{p}\| \leq r\}$$

Definition 3. *Given a subset $A \subset \mathbb{Z}^2$, the ζ-dimension of A, written $Dim_\zeta(A)$, is defined as*

$$Dim_\zeta(A) =_{def} \limsup_{r \to \infty} \frac{\log\left|A \cap B_r\left(\vec{0}\right)\right|}{\log r}.$$

This definition is not the typical definition of ζ-dimension, but it is equivalent [5]. Another equivalent definition for ζ-dimension that will be more useful in some of our cases is

$$Dim_\zeta(A) = \limsup_{r \to \infty} \frac{\log \left| A \cap B_{2^r}\left(\vec{0}\right) \right|}{r}$$

where we consider balls of radius 2^r rather than just r.

Discrete Self-similar Fractals. Having introduced ζ-dimension, we now define the class of discrete self-similar fractals. Because of their discrete nature, a consequence of their bottom-up recursive definition, this class of fractals is particularly suited to be investigated with respect to the aTAM, which is inherently confined to the integer lattice.

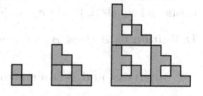

Fig. 3. The first 3 stages of the Sierpinski triangle DSSF.

Definition 4. *Let $G = \{g_0, \ldots, g_{k-1}\}$ be a non-empty, finite shape in $\mathbb{N}^2$ satisfying the following where w is the largest x coordinate of a point in G and h is the largest y coordinate:*

- *There exist two points $(x_0, y_0), (x_1, y_1) \in G$ such that $x_0 = 0$ and $y_1 = 0$,*
- *There exist two points $(x_0, y_0), (x_1, y_1) \in G$ such that $x_0 = 0$, $x_1 = w$ and $y_0 = y_1$,*
- *There exists two points $(x_0, y_0), (x_1, y_1) \in G$ such that $y_0 = 0$, $y_1 = h$ and $x_0 = x_1$, and*
- *there exists at least one point $(x, y) \in \mathbb{N}^2$ with $0 \le x \le w$ and $0 \le y \le h$ such that (x, y) is not in G.*

Define $F_0 = G$ and, for $i > 0$, define

$$F_i = \left\{ F_{i-1} + \left(w^i \cdot x, h^i \cdot y\right) : (x, y) \in G \right\}$$

where a set of points S plus a point (x, y) refers to translation of each point in S by the vector corresponding to (x, y). Each F_i is called a stage of the fractal.

$$F = \bigcup_{i \in \mathbb{N}} F_i$$

is then called a discrete self-similar fractal with generator G.

Intuitively, discrete self-similar fractals can be thought of as being constructed by iteratively replacing each point in a stage by an entire copy of the generator. The conditions for generators assure that the resulting DSSF is connected and not degenerate (i.e. just a point, line, or full quadrant). (Several examples can be found in [17,20]).

3   Sparsity

In this section, we introduce the concept of sparsity and prove some results which relate sparsity to ζ-dimension.

Definition 5. *Given a shape* $A \subset \mathbb{Z}^2$*, let*

$$\partial_A(r) =_{def} \sup_{\vec{p} \in \mathbb{Z}^2} \frac{\left| A \cap B_r\left(\vec{p}\right) \right|}{\left| B_r\left(\vec{p}\right) \right|}$$

We call $\partial_A(r)$ the *radius-r partial density* of A. Additionally, if $\lim_{r \to \infty} \partial_A(r)$ exists and equals d, then we say that A has *density d*.

Definition 6. *A shape* A *whose density exists and equals* 0 *is called* sparse.

3.1   Sparsity and ζ-Dimension

Here we show that for sparse shapes, if the radius r density can be bounded above or below by a sufficiently well behaved function, then we can put a bound on its ζ-dimension. Informally, this shows that ζ-dimension can be thought of in terms of the rate at which the partial densities of a shape vanish as the radius increases.

Theorem 1. *If* $\partial_A(r) = O\left(r^{-k}\right)$*, then* $Dim_\zeta(A) \le 2 - k$.

Proof. Let A be a shape such that $\partial_A(r) = O\left(r^{-k}\right)$, that is to say there exists R and c such that for all $r > R$,

$$\sup_{\vec{p} \in \mathbb{Z}^2} \frac{\left| A \cap B_r\left(\vec{p}\right) \right|}{\left| B_r\left(\vec{p}\right) \right|} \le cr^{-k}.$$

Since $\vec{0} \in \mathbb{Z}^2$, this implies that

$$\frac{\left| A \cap B_r\left(\vec{0}\right) \right|}{\left| B_r\left(\vec{0}\right) \right|} \le cr^{-k} \implies \left| A \cap B_r\left(\vec{0}\right) \right| \le cr^{-k} \left| B_r\left(\vec{0}\right) \right| \le c'r^{2-k}$$

where c' is some constant independent of r and R, since $\left| B_r\left(\vec{0}\right) \right| = \Theta\left(r^2\right)$. By definition

$$Dim_\zeta(A) = \limsup_{r \to \infty} \frac{\log \left| A \cap B_r\left(\vec{0}\right) \right|}{\log r},$$

and thus

$$Dim_\zeta(A) \le \limsup_{r \to \infty} \frac{\log \left(c'r^{2-k}\right)}{\log r} = \limsup_{r \to \infty} \frac{\log c' + \log \left(r^{2-k}\right)}{\log r} = 2 - k.$$

$\square$

Theorem 2. *If $\partial_A(r) = \Omega\left(r^{-k}\right)$, then $Dim_\zeta(A) \geq 2 - k$.*

The proof for this theorem is nearly identical to the previous proof and the following corollary follows immediately from both of these theorems.

Corollary 1. *If $\partial_A(r) = \Theta\left(r^{-k}\right)$, then $Dim_\zeta(A) = 2 - k$.*

3.2 Sparsity of Discrete Self-similar Fractals

We have just shown how the concept of sparsity is related to the concept of ζ-dimension. Still, it is not immediately obvious that sparsity is a useful property to consider. Here we prove 2 results which demonstrate the concept's utility.

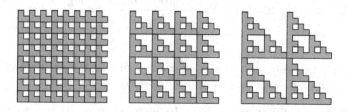

Fig. 4. The shapes A_i used to show that DSSFs are sparse, are infinitely repeating grids of a single stage of a fractal F_i.

Theorem 3. *Discrete self-similar fractals are sparse.*

Proof. Let A be a DSSF generated by G and let w be the largest x coordinate of a point in G, and h be the largest y coordinate. Additionally, let $F_0, F_1, F_2, \ldots$ be the stages of A, which are shapes themselves, and let $A_0, A_1, A_2, \ldots$ be the shapes consisting of the plane tiled by copies of $F_0, F_1, F_2, \ldots$ respectively. These are illustrated in Fig. 4. It's not difficult to see that the density of A_i will be $\frac{|G|^i}{(wh)^i}$. Since $|G| < wh$, these densities converge to 0 as we consider bigger and bigger stages. Since A is a subset of each A_i, the density of A cannot be bigger than the density of any A_i. Thus, since there is an A_i with density smaller than any positive integer, A must have density 0.   □

Theorem 4. *If A is a DSSF with generator G and if the largest x and y coordinate of points in G are the same (i.e. G is bounded by a square), then $Dim_\zeta(A) = \frac{\log|G|}{\log s}$ where s is the largest x and y coordinate of a point in G.*

Proof. Let s be the largest x and y coordinate of a point in G and let $g = \frac{|G|}{s^2}$. Consider what happens to $\partial_A(r)$ when $r = s^i$ for any $i \in \mathbb{N}$. At this scale, $\partial_A(r)$ is largely determined by the ith stage of A. It's certain that $\partial_A(r) \geq g^i$ since an entire copy of the ith stage of A could fit inside a ball of radius r. Additionally $\partial_A(r)$ is no bigger than $9g^i$ since the ball will always fit inside at most 9 copies of the ith stage of the fractal arranged in a square. Therefore $\partial_A(r) = \Theta\left(g^i\right)$ when $r = s^i$.

Let $k = -\log_s(g)$. Then notice that if $r = s^i$,

$$r^{-k} = s^{-ik} = s^{i \log_s(g)} = g^i$$

so since $\partial_A(r) = \Theta(g^i)$, it must be the case that $\partial_A(r) = \Theta(r^{-k})$ for $r = s^i$. Additionally notice that for all r, $\partial_A(r)$ is monotonically decreasing since, as we consider larger and larger radii, the numerator in the definition of $\partial_A(r)$ will never increase by a number larger than the denominator. Given that $\partial_A(r)$ is monotonically decreasing and since on a subsequence where $r = s^i$, $\partial_A(r) = \Theta(r^{-k})$, it must be the case that for all r, $\partial_A(r) = \Theta(r^{-k})$. Therefore by Corollary 1, $\text{Dim}_\zeta(A)$ must be $2 - k$.

$$2 - k = 2 - \log_s(g) = \frac{2\log(s) - \log(g)}{\log(s)} = \frac{\log|G|}{\log s}$$

$\square$

Theorem 5. *Any shape which is producible by an aTAM system which requires standard[1] counter gadgets that count arbitrarily far is not sparse.*

Proof. Because counter gadgets have to increase in width to accommodate increasing counter values, for any radius r, there will be some point $\vec{p}$ inside the counter gadget after some finite counter value such that $B_r(\vec{p})$ fits entirely within the tiles of the counter. This means that the partial densities for all radii will be 1 and thus that the produced shape is not sparse. $\square$

In [17,20], systems were presented that are capable of self-assembling shapes that approximate fractals with a technique called "fibering", and in [15,18] the fractal-approximation is "laced". These approximations have the same general shapes as their discrete self-similar fractal counterparts, and the same ζ-dimensions. However, they also utilize standard counter gadgets so that Theorem 5 shows that they are not sparse.

4   Arbitrary Fractal Dimension Construction

In this section, we show our main positive result, which is that it is possible to uniquely self-assemble a shape whose ζ-dimension is equal to a pre-specified algebraic number.

Theorem 6. *There exists an aTAM tileset T such that, for all $x \in \mathbb{R}$ satisfying:*

- *$1 < x < 2$, and*
- *there exists a deterministic Turing machine which halts in $\Theta\left(2^{\frac{xn}{4}}\right)$ steps given input n encoded in binary,*

---

[1] Here a *standard* counter gadget refers to commonly used log-width counter gadgets. It is unknown whether or not counter-like gadgets can be implemented in a sparse way.

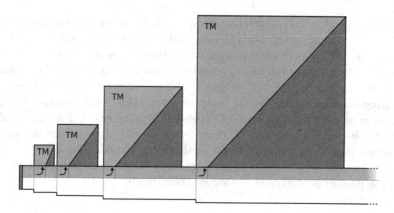

Fig. 5. Illustration of the arbitrary fractal dimension construction. The planter at the bottom propagates and rotates a Turing machine description and the width (number of digits) of the counter (illustrated in gray along the planter) to be used as inputs for Turing machine gadgets which grow upwards increasing in width to account for additional space. Dark Gray tiles indicate filler tiles that "square up" each TM gadget once it has finished its computation.

there exists a directed TAS $\mathcal{T}_x = (T, \sigma_x, 2)$ whose only terminal assembly has ζ-dimension x.

To prove this theorem, given an x satisfying the above constraints with corresponding Turing machine M, we describe the directed TAS $\mathcal{T}_x = (T, \sigma_x, 2)$ whose terminal assembly has ζ-dimension x. Note that our construction only requires that the seed σ_x depend on the value of x; the tileset T and the temperature $\tau = 2$ remain constant over all x so that T is universal for this class of systems.

4.1 Construction Description

Here we describe the construction at a high enough level that anyone familiar with common tile-assembly gadgets would easily be able to work out the tile-level details. This construction has 2 components illustrated in Fig. 5 which we call the *planter* and the *Turing machine array*.

The Planter. The planter is a standard planter gadget used in many aTAM constructions [13, 16] and consists of a log-width counter gadget which, in addition to keeping track of a binary counter value, two other values are propagated along the counter: the width (number of digits) of the counter and the description of our Turing machine M. We will use the symbol ν to index the *counter value* in each column of our planter and denote the *counter width* with $\lambda(\nu) = \lceil \log(\nu) \rceil$ (or just λ if ν is obvious in context) and the Turing machine description with μ. The initial counter value ν_0 and TM description μ are provided as input to our TAS $\mathcal{T}$ in the form of the seed σ_x which acts as the first column of our planter.

We choose ν_0 so that it is large enough for all of the necessary arguments to be made in the following proof, but the exact value is not important.

Each time the counter increases its width (when ν is a power of 2), special tiles attach which rotate the values μ and $\lambda(\nu)$ northward. We call the columns where this rotation occurs *TM columns*. These values are then used as input to a Turing machine in the Turing machine array. This rotation along with the propagation of a constant number of values in the planter occur using standard gadgets in the aTAM and can be done using special glues inside the planter construction so that the overall shape is that of a standard log-width counter. Also note that the width of the planter increases to the south so that the north side of the planter is always at the same y coordinate.

The Turing Machine Array. At each TM column, the values μ and λ are rotated to the north and used as input for a standard Turing machine gadget. This gadget is hard-coded to simulate the behavior of a fixed Turing machine (independent of x). In our case we choose this fixed Turing machine to be a deterministic universal Turing machine U which simulates the Turing machine described by the input μ on the string encoding the input λ. For ease of analysis, we choose U so that the amount of time it takes to simulate M induces a fixed quadratic blowup in runtime[2]. That is, if on input w, M runs in time $T(|w|)$, then U on input (μ, w) runs in time $\Theta\left(T(|w|)^2\right)$.

There are many permutations of Turing machine gadgets used in all sorts of aTAM constructions. For our purposes, we choose tiles so that the number of rows and columns in the completed gadget are exactly equal to the number of steps used by the Turing machine U on the given input, resulting in a square of tiles. This is not hard to do using standard aTAM gadgets. Since we required that M runs in time $\Theta\left(2^{xn/4}\right)$ on input n encoded in binary and that U induces a quadratic blowup in runtime, the side length of this square of tiles will be $\Theta\left(2^{\frac{xn}{2}}\right)$ and the total number of tiles in each gadget will be $\Theta\left(2^{xn}\right)$. Also note that since $x < 2$ by definition, the side length of our Turing machine gadgets is small enough that adjacent Turing machine gadgets will never collide supposing that our initial counter value is chosen to be sufficiently large. For convenience we also index the Turing machine gadgets by the value of their input so that T_λ is the Turing machine gadget which received input value λ.

4.2   Proof of Correct Fractal Dimension

Here we show that the terminal assembly of $\mathcal{T}_x$ has fractal dimension x.

Proof. Let x be defined as above and let α be the terminal assembly of $\mathcal{T}_x$ and let A be the shape of α. We can define A as the disjoint union of two shapes A_p and A_t where A_p is the set of tile locations corresponding to tiles in the planter and A_t

---

[2] There are universal Turing machines which induce asymptotically smaller runtime blowups, but choosing one with a quadratic blow up makes analysis of the final fractal dimension easier.

is the set of tile locations corresponding to tiles in the Turing machine array. Since A is the disjoint union of A_p and A_t, $\text{Dim}_\zeta(A) = \max\{\text{Dim}_\zeta(A_p), \text{Dim}_\zeta(A_t)\}$ [5]. Therefore it suffices to compute the ζ-dimensions of these subsets. Additionally, since ζ-dimension is invariant under any bi-Lipschitz transformation of $\mathbb{Z}^2$ [5] (which includes any constant translation), we can choose any orientation or translation that is convenient.

Since by definition the planter takes the shape of a log-width counter gadget initialized to some value, A_p is a subset of the shape, which we will call A_c, of a log-width counter starting at value 1. A_c only has a constant number of columns that A_p does not. We can choose to define A_c so that the value encoded in a column of the counter plus 1 is the x coordinate of that column in $\mathbb{Z}^2$ and the one's place row of the counter is at y coordinate 0. By this convention, the width of the counter increases at x coordinates corresponding to a power of 2 plus 1 and it is not difficult to see that $A_c \cap B_{2^r}\left(\vec{0}\right) = \Theta(r2^r)$. Therefore $\text{Dim}_\zeta(A_c) = \limsup_{r\to\infty} \frac{\log(\Theta(r2^r))}{r} = 1$. Since A_p only differs by a constant number of tile locations from A_c up to translation, $\text{Dim}_\zeta(A_p) = 1$ as well.

Now we consider the Turing machine array. For convenience, we choose the translation of A_t so that westernmost tiles in each Turing machine gadget square are at the x coordinate equal to its input λ plus 1. This is simply a translation and requires no scaling because the rotation of inputs into the Turing machine gadgets occurred when the planter increased its counter width, which occurs with identical spacing. Additionally, we choose our translation of A_t so that the southernmost tiles in each gadget sit at y coordinate 0.

By this convention it is easy to see that the $B_{2^r}\left(\vec{0}\right)$ will only contain tiles from gadgets T_λ with $\lambda < r$. Additionally, while it may be the case that tiles from T_r sit outside of $B_{2^r}\left(\vec{0}\right)$, since the corner of the square of tiles may extend past the ball, this only occurs a finite number of times. This is because the side length of the square is $\Theta\left(2^{\frac{xn}{2}}\right)$ where $x < 2$ and so the side length is asymptotically smaller than $\Theta(2^n)$ which is the spacing between adjacent gadgets. After some finite number of gadgets, the side length will eventually be small enough to completely fit inside of the corresponding ball.

Since gadget T_n has $\Theta(2^{nx})$ tiles, and a sum of a finite number of consecutive powers of 2 is the next power of 2 minus 1, given that r is sufficiently large, it is easy to see that

$$\left|A_t \cap B_{2^r}\left(\vec{0}\right)\right| = \Theta(2^{rx}).$$

Therefore

$$\text{Dim}_\zeta(A_t) = \limsup_{r\to\infty} \frac{\log(\Theta(2^{rx}))}{r} = x.$$

Since $x > 1$ by definition, this means that $\text{Dim}_\zeta(A_p \cup A_t) = x$ which means identically that $\text{Dim}_\zeta(A) = x$. □

Corollary 2. *The set of ζ-dimensions that satisfy the constraint of the previous theorem includes all algebraic numbers.*

This follows from the fact that the digits of algebraic numbers are computable in polylogarithmic time [9]. Because of this we can define our Turing machine to first compute the non-fractional digits of nx and then use that value as the number of bits for a counter. Our Turing machine will begin counting and then halt once the counter reaches 2^{nx}. Additionally, the converse of this corollary is not true and the class of achievable fractal dimensions includes some transcendental numbers as well since constants like $\pi - 2$ and $e - 1$ can be used.

5   Conclusion

In this paper, we investigated the extent to which tile assembly systems in the aTAM are able to self-assemble shapes having fractal dimensions between 1 and 2. To that end, we first introduced a concept of sparsity, which we used as a tool for investigating such systems, proving a relationship between it and the ζ dimension of discrete self-similar fractals. Then, we gave a construction showing that it is possible to uniquely self-assemble a shape whose ζ-dimension is equal to a pre-specified algebraic number. Thus, we have presented some incremental results towards the following conjecture (see also [20, 22]): No non-trivial discrete self-similar strictly self-assembles in the aTAM.

Acknowledgments. The authors would like to thank the three anonymous reviewers whose comments helped improve the presentation and technical correctness of this paper.

References

1. Barth, K., Furcy, D., Summers, S.M., Totzke, P.: Scaled tree fractals do not strictly self-assemble. In: Ibarra, O.H., Kari, L., Kopecki, S. (eds.) UCNC 2014. LNCS, vol. 8553, pp. 27–39. Springer, Cham (2014). https://doi.org/10.1007/978-3-319-08123-6_3
2. Cannon, S., et al.: Two hands are better than one (up to constant factors): self-assembly in the 2HAM vs. aTAM. In: Portier, N., Wilke, T. (eds.) STACS, volume 20 of LIPIcs, pp. 172–184. Schloss Dagstuhl - Leibniz-Zentrum fuer Informatik (2013)
3. Chalk, C.T., Fernandez, D.A., Huerta, A., Maldonado, M.A., Schweller, R.T., Sweet, L.: Strict self-assembly of fractals using multiple hands. Algorithmica **76**(1), 1–30 (2015). https://doi.org/10.1007/s00453-015-0022-x
4. Demaine, E.D., Patitz, M.J., Rogers, T.A., Schweller, R.T., Summers, S.M., Woods, D.: The two-handed tile assembly model is not intrinsically universal. In: Fomin, F.V., Freivalds, R., Kwiatkowska, M., Peleg, D. (eds.) ICALP 2013. LNCS, vol. 7965, pp. 400–412. Springer, Heidelberg (2013). https://doi.org/10.1007/978-3-642-39206-1_34
5. Doty, D., Gu, X., Lutz, J.H., Mayordomo, E., Moser, P.: Zeta-Dimension. In: Jędrzejowicz, J., Szepietowski, A. (eds.) MFCS 2005. LNCS, vol. 3618, pp. 283–294. Springer, Heidelberg (2005). https://doi.org/10.1007/11549345_25

6. Doty, D., Lutz, J.H., Patitz, M.J., Schweller, R.T., Summers, S.M., Woods, D.: The tile assembly model is intrinsically universal. In: Proceedings of the 53rd Annual IEEE Symposium on Foundations of Computer Science. FOCS 2012, pp. 302–310 (2012)
7. Evans, C.G.: Crystals that count! Physical principles and experimental investigations of DNA tile self-assembly. Ph.D. thesis, California Institute of Technology (2014)
8. Furcy, D., Summers, S.M.: Scaled pier fractals do not strictly self-assemble. Nat. Comput. 16(2), 317–338 (2015). https://doi.org/10.1007/s11047-015-9528-z
9. Hartmanis, J., Stearns, R.E.: On the computational complexity of algorithms. Trans. Am. Math. Soc. 117, 285–306 (1965)
10. Hendricks, J., Opseth, J., Patitz, M.J., Summers, S.M.: Hierarchical growth is necessary and (sometimes) sufficient to self-assemble discrete self-similar fractals. In: Doty, D., Dietz, H. (eds.) DNA 2018. LNCS, vol. 11145, pp. 87–104. Springer, Cham (2018). https://doi.org/10.1007/978-3-030-00030-1_6
11. Hendricks, J., Olsen, M., Patitz, M.J., Rogers, T.A., Thomas, H.: Hierarchical self-assembly of fractals with signal-passing tiles (extended abstract). In: Rondelez, Y., Woods, D. (eds.) DNA 2016. LNCS, vol. 9818, pp. 82–97. Springer, Cham (2016). https://doi.org/10.1007/978-3-319-43994-5_6
12. Hendricks, J., Opseth, J.: Self-assembly of 4-sided fractals in the two-handed tile assembly model. In: Patitz, M.J., Stannett, M. (eds.) UCNC 2017. LNCS, vol. 10240, pp. 113–128. Springer, Cham (2017). https://doi.org/10.1007/978-3-319-58187-3_9
13. Hendricks, J., Patitz, M.J., Rogers, T.A.: Universal simulation of directed systems in the abstract tile assembly model requires undirectedness. In: Proceedings of the 57th Annual IEEE Symposium on Foundations of Computer Science (FOCS 2016), New Brunswick, New Jersey, USA 9–11 October 2016, pp. 800–809 (2016)
14. Kautz, S.M., Lathrop, J.I.: Self-assembly of the discrete Sierpinski carpet and related fractals. In: Deaton, R., Suyama, A. (eds.) DNA 2009. LNCS, vol. 5877, pp. 78–87. Springer, Heidelberg (2009). https://doi.org/10.1007/978-3-642-10604-0_8
15. Kautz, S.M., Shutters, B.: Self-assembling rulers for approximating generalized Sierpinski carpets. Algorithmica 67(2), 207–233 (2013)
16. Lathrop, J.I., Lutz, J.H., Patitz, M.J., Summers, S.M.: Computability and complexity in self-assembly. Theory Comput. Syst. 48(3), 617–647 (2011)
17. Lathrop, J.I., Lutz, J.H., Summers, S.M.: Strict self-assembly of discrete Sierpinski triangles. Theoret. Comput. Sci. 410, 384–405 (2009)
18. Lutz, J.H., Shutters, B.: Approximate self-assembly of the Sierpinski triangle. Theory Comput. Syst. 51(3), 372–400 (2012)
19. Padilla, J.E., Patitz, M.J., Schweller, R.T., Seeman, N.C., Summers, S.M., Zhong, X.: Asynchronous signal passing for tile self-assembly: fuel efficient computation and efficient assembly of shapes. Int. J. Found. Comput. Sci. 25(4), 459–488 (2014)
20. Patitz, M.J., Summers, S.M.: Self-assembly of discrete self-similar fractals. Nat. Comput. 1, 135–172 (2010)
21. Patitz, M.J., Summers, S.M.: Self-assembly of decidable sets. Nat. Comput. 10(2), 853–877 (2011)
22. Patitz, M.J., Summers, S.M.: Self-assembly of infinite structures: a survey. Theor. Comput. Sci. 412(1-2), 159–165 (2011). https://doi.org/10.1016/j.tcs.2010.08.015
23. Rothemund, P.W.K., Papadakis, N., Winfree, E.: Algorithmic self-assembly of DNA Sierpinski triangles. PLoS Biol. 2(12), e424 (2004)

24. Rothemund, P.W.K., Winfree, E.: The program-size complexity of self-assembled squares (extended abstract). In: STOC 2000: Proceedings of the Thirty-second Annual ACM Symposium on Theory of Computing, pp. 459–468, Portland, Oregon, United States. ACM (2000)
25. Soloveichik, D., Winfree, E.: Complexity of self-assembled shapes. SIAM J. Comput. **36**(6), 1544–1569 (2007)
26. Winfree, E.: Algorithmic Self-Assembly of DNA. Ph.D. thesis, California Institute of Technology, June 1998
27. Woods, D., et al.: Diverse and robust molecular algorithms using reprogrammable DNA self-assembly. Nature **567**, 366–372 (2019)

Bistable Latch Ising Machines

Jaijeet Roychowdhury[(✉)]

Department of Electrical Engineering and Computer Sciences,
University of California, Berkeley, CA 94720, USA
jr@berkeley.edu

Abstract. Ising machines have been attracting attention due to their ability to use mixed discrete/continuous mechanisms to solve difficult combinatorial optimization problems. We present BLIM, a novel Ising machine scheme that uses latches (bistable elements) with controllable gains as Ising spins. We show that networks of coupled latches have a Lyapunov or "energy" function that matches the Ising Hamiltonian in discrete operation, enabling them to function as Ising machines. This result is established in a general coupled-element Ising machine framework that is not limited to BLIM. Operating the latches periodically in analog/continuous mode, during which bistability is removed, helps the system traverse to better minima. CMOS realizations of BLIM have desirable practical features; implementation in other physical domains is an intriguing possibility.

1 Introduction

Over the last decade, hardware Ising machines have emerged as a promising means to solve classically difficult (*e.g.*, NP-complete) computational problems. The premise of Ising machines is that specialized hardware implementing the Ising computational model (see Sect. 2) can solve difficult combinatorial problems more effectively than classical algorithms (such as semidefinite programming and simulated annealing [7,10]) run on digital computers. Ising machines first came into prominence with the D-Wave quantum annealer [2,8] and the Coherent Ising Machine (CIM) [14,18,19]. A D-Wave quantum annealer with 5000 spins is available commercially; CIM with 2000 spins has been successfully demonstrated at NTT Research Labs, with larger systems under active development. Although they have established the field of Ising machines and inspired follow-on technologies, D-Wave's quantum annealer and CIM are physically large, expensive, and difficult to miniaturize or scale to larger problems. A few years ago, we showed that networks of coupled oscillators can be designed to function as Ising machines [3,15–17]. Such oscillator Ising machines (OIMs) changed the technology landscape for Ising machines by bringing them within the realm of miniaturizable CMOS electronics, with all the size, cost, speed, energy efficiency, scalability and mass production benefits that accrue as a result.

We now present the Bistable Latch Ising Machine (BLIM), a new way to build Ising machines that employs simple bistable elements, *i.e.*, latches, a familiar

I. Kostitsyna and P. Orponen (Eds.): UCNC 2021, LNCS 12984, pp. 131–148, 2021.
https://doi.org/10.1007/978-3-030-87993-8_9

and ubiquitous element in electronics.[1] BLIM is enabled by a result (Sect. 3) that establishes that networks of coupled latches have an "energy[2] function" that is naturally minimized, leading to good solutions of the Ising problem. That latches can be used as substrates represents a broadening, both theoretical and practical, of the Ising machine landscape, while enhancing the advantages of miniaturizability/scalability, low cost, and mass production introduced by OIM. Latches are simpler elements than oscillators, with basic versions requiring only 4 CMOS transistors. Importantly, the formulation in which we prove our main result is a general one, not limited to latches—it encompasses OIM and, potentially, other types of Ising machines. Using this formulation to explore and compare the operational mechanisms of BLIM and OIM may lead to progress on a central question: how exactly do Ising machines work?

The remainder of the paper is organized as follows. In Sect. 2, we provide background on Ising models, oscillator Ising machines and latches. In Sect. 3, we set up a suitable system of equations for coupled latches, abstract them to a generalized form, and prove our main result: that the system has a Lyapunov function that matches a corresponding Ising Hamiltonian for high values of latch gain. Illustrative examples are provided in Sect. 4.

2   Background

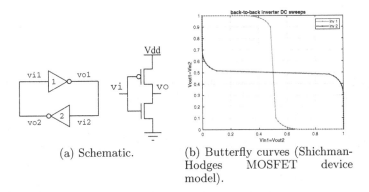

(a) Schematic.   (b) Butterfly curves (Shichman-Hodges   MOSFET   device model).

Fig. 1. Back to back inverters implement a bistable latch.

2.1   The Ising Model

The Ising model is simply a weighted graph, *i.e.*, a collection of nodes/vertices and branches/edges between some pairs of nodes, with each branch having a real-number weight. Each node (termed a "spin" in this context) is allowed to

---

[1] BLIM is not limited to electronic latches; it can use latches from any domain, *e.g.*, biochemical latches [5,6].

[2] This "energy" is not obviously related to any concept of physical energy, which latches, like all practical electronic elements, consume and dissipate as heat.

take two values, either 1 or −1. Associated with this graph is an expression, the **Ising Hamiltonian**, which multiplies the weight of each branch by the values of the two spins it connects to, and sums over all branches, *i.e.*,

$$H = -\frac{1}{2} \sum_{i,j=1}^{N} J_{ij}\, s_i s_j, \quad \text{where } J_{ij} = J_{ji},\ J_{ii} = 0,\ \text{and } s_i \in \{-1,\ +1\} \quad (1)$$

are the N spins. J_{ij} are the branch weights, also called coupling coefficients. Owing to the Ising problem's origins for modelling and explaining ferromagnetism [4], Ising Hamiltonians are sometimes interpreted as an "energy" associated with a given configuration of the spins, although in recent computational applications they usually have no connection with energy in physics. The "Ising problem" is to find spin configurations with the minimum possible energy.

2.2 Latches

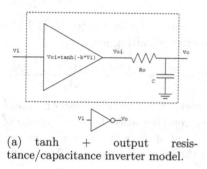

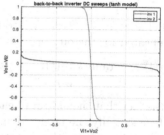

(a) tanh + output resistance/capacitance inverter model.

(b) Latch butterfly curves (tanh inverter model).

Fig. 2. tanh() + output resistance/capacitance inverter model and corresponding latch butterfly curves.

A fundamental element in electronics, the latch is perhaps most easily understood as two back-to-back inverters, as shown in Fig. 1(a), with each inverter consisting of the 2-transistor circuit shown at the right. Latches are ubiquitous in digital systems, in which they are the basis for, *e.g.*, registers and SRAM (static random access memory); as such, they are among the most compact and power-efficient elements in CMOS electronics.[3] That they are bistable becomes apparent when the I/O curves of both inverters are depicted on the same axes to produce so-called butterfly curves, shown in Fig. 1(b) (generated using the Shichman-Hodges model [13] for the MOSFETs).

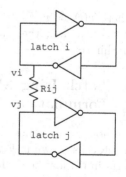

Fig. 3. Coupling between latches i and j.

---

[3] When not switching, CMOS consumes no power beyond leakage losses.

The three intersections are the DC solutions (*i.e.*, equilibria) of the circuit. The intersection in the middle can be shown to be dynamically unstable,[4] leaving two stable solutions. The nature of the curves in Fig. 1(b), and the intersections that lead to bistability, are the essential feature of any latch, be it electronic or from any other domain (*e.g.*, biological [5,6]).

For concreteness and ease of exposition, we abstract each inverter using a tanh($\cdot$) voltage transfer characteristic[5] followed by an output resistance and load capacitor, as shown in Fig. 2(a). The voltage I/O characteristic is $v_{oi} = \tanh(-kv_i)$, where the parameter $k > 0$ controls the gain, or sharpness, of the inverter characteristic. No current is drawn at the input; in the absence of loading at the output, the output voltage $v_o = v_{oi}$ (at DC). The corresponding butterfly curves, for $k = 20$, are shown in Fig. 2(b)—note their similarity to the ones in Fig. 1(b) for CMOS inverters.

The capacitor in Fig. 2(a) introduces dynamics, resulting in the following differential equation for a single inverter in the absence of additional load at v_o:

$$C\frac{dv_o(t)}{dt} = \frac{\tanh(-kv_i(t)) - v_o(t)}{R_o} = -\frac{\tanh(kv_i(t)) + v_o(t)}{R_o}. \tag{2}$$

To model a latch, *i.e.*, two back-to-back inverters are connected as in Fig. 1(a). (2) is repeated for the output of each inverter, resulting in

$$C\frac{dv_{i1}(t)}{dt} = -\frac{\tanh(kv_{i2}(t)) + v_{i1}(t)}{R_o}, \quad C\frac{dv_{i2}(t)}{dt} = -\frac{\tanh(kv_{i1}(t)) + v_{i2}(t)}{R_o}, \tag{3}$$

where the fact that the output of each inverter is the input of the other has been used. If we simplify the latch's dynamical representation by ignoring one of the inverter capacitors (*e.g.*, the one at the output of the first inverter $v_{o1} = v_{i2}$; this does not sacrifice any essential aspect of the latch's operation), (3) simplifies to the single differential equation

$$C\frac{dv(t)}{dt} = -\frac{\tanh\left(-k\tanh(kv(t))\right) + v(t)}{R_o} = \frac{\tanh\left(k\tanh(kv(t))\right) - v(t)}{R_o}, \tag{4}$$

where $v(t) \triangleq v_{i1}(t)$. (4) is the starting point for establishing the key result in Sect. 3, *i.e.*, that interconnected systems of latches can function as Ising machines.

3   Latch Ising Machines: A General Lyapunov Formulation

We now set up a network of coupled latches. Each coupling is realized using a resistor, as shown in Fig. 3. The coupling from the jth latch appears as an extra term in (4) for the ith latch, whose equation becomes

---

[4] *i.e.*, any small perturbation (*e.g.*, due to noise) from this solution will make the latch settle to one of the other two solutions, at the top left and bottom right.

[5] tanh() is merely a convenient analytical choice; any other smoothed step-like function can be used instead.

$$C\frac{dv_i(t)}{dt} = \frac{\tanh\left(k\tanh(kv_i(t))\right) - v_i(t)}{R_o} - \frac{v_i - v_j}{R_{ij}}$$

$$= G\tanh\left(k\tanh(kv_i(t))\right) - Gv_i(t) - J_{ij}(v_i - v_j), \tag{5}$$

where $G \triangleq \frac{1}{R_o}$ and $J_{ij} \triangleq \frac{1}{R_{ij}}$. For N latches, this becomes a system of N differential equations:

$$\frac{dv_i(t)}{dt} = \frac{1}{C}\left[G\tanh\left(k\tanh(kv_i(t))\right) - Gv_i(t) - \sum_{j=1}^{N} J_{ij}(v_i - v_j)\right], \tag{6}$$

$$i = 1, \cdots, N.$$

(6) can be written in a more general form, as

$$\frac{dv_i(t)}{dt} = f(v_i; k) - \sum_{j=1}^{N} J_{ij}\, g(v_i, v_j; k), \quad i = 1, \cdots, N; \tag{7}$$

choosing

$$f(v_i; k) = \frac{G}{C}\left(\tanh\left(k\tanh(kv_i)\right) - v_i\right) \text{ and } g(v_i, v_j; k) = \frac{v_i - v_j}{C} \tag{8}$$

turns it into (6). The utility of (7) over (6) is its generality: coupled networks of any kind of latch can be represented by appropriate choice of $f(\cdot; \cdot)$ and $g(\cdot, \cdot; \cdot)$. Indeed, (7) is not limited to latch networks; *e.g.*, OIM using the the Kuromoto model with SHIL [15], for coupled oscillator systems, is also captured, by setting

$$f(\Delta\phi_i; A_s) = \frac{A_s}{w_0}\sin\left(2\Delta\phi_i\right) \text{ and } g(\Delta\phi_i, \Delta\phi_j; A_s) = -\frac{A_c}{w_0}\sin\left(\Delta\phi_i - \Delta\phi_j\right). \tag{9}$$

The development in the remainder of this section does not use the specific form of the $\tanh(\cdot)$ latch model (6); instead, it uses the more general form of (7), thereby being applicable to different latch models, as well as OIM and potentially other manifestations of Ising machines. We prove two key results: 1) that there is a Lyapunov function (11) for (7),[6] and 2) that for high values of k, at which latches exhibit bistability, the Lyapunov function matches the Ising Hamiltonian (Theorem 2). In other words, the same underlying properties that enable coupled oscillator systems to serve as Ising machines hold for coupled latch systems.

3.1   Lyapunov Function

As already noted, the coupling is assumed to be symmetric, with no "self coupling",[7] *i.e.*,

---

[6] The Lyapunov function is defined in terms of abstract functions $z(\cdot; \cdot)$ and $h(\cdot, \cdot; \cdot)$ that are related to the functions $f(\cdot; \cdot)$ and $g(\cdot, \cdot; \cdot)$ in the generalized model (7). The relations, captured abstractly as assumptions in Assumption 2, are illustrated concretely for the $\tanh(\cdot)$ latch model in Sect. 4.

[7] This assumption is intrinsic to the Ising model, as already noted in (1).

Assumption 1 (Coupling properties)

$$J_{ij} = J_{ji}, \quad J_{ii} = 0, \quad i, j = 1, \cdots, N. \tag{10}$$

We now define a scalar function of the $\{v_i\}$ that we will show satisfies the properties of a Lyapunov function. The functions $z(\cdot; \cdot)$ and $h(\cdot, \cdot; \cdot)$ used in the definition are left abstract at this point; specific choices for BLIM will be made in (39) and (41), later.

Definition 1 (Lyapunov function $L(\cdots)$). *Define*

$$L(v_1, \cdots, v_N; k) \triangleq -\sum_{i=1}^{N} \left(z(v_i; k) - \sum_{j=1}^{N} J_{ij} h(v_i, v_j; k) \right). \tag{11}$$

Denoting $\vec{v} \triangleq [v_1, \cdots, v_N]^T$, *we will also write this as* $L(\vec{v}; k)$. $z(v; k)$ *and* $h(v_1, v_2; k)$ *are continuous and differentiable functions with properties to be stated later. Hence* $L(\vec{v}; k)$ *is continuous and differentiable.*

We now assume that $z(\cdot; \cdot)$ and $h(\cdot, \cdot; \cdot)$ in (11) satisfy the following properties and relations to $f(\cdot; \cdot)$ and $g(\cdot, \cdot; \cdot)$. The first assumption (12) is essentially a definition of $z(\cdot; \cdot)$, used in the Lyapunov function, in terms of the abstraction $f(\cdot; \cdot)$ of the tanh$(\cdot)$ latch model, used in (7). The second assumption captures the essential relation between the abstracted coupling function $g(\cdot, \cdot; \cdot)$ in (7), and the corresponding function $h(\cdot, \cdot; \cdot)$ in the Lyapunov expression in (11). This relation is required in order to show (in Theorem 1, below) that (11) is indeed a Lyapunov function for (7).

Assumption 2 (Properties of $z(\cdot; \cdot)$ and $h(\cdot, \cdot; \cdot)$)

1. $f(\cdot)$ in (8) is the derivative of $z(\cdot)$ in (11):

$$f(v_m; k) = \frac{dz(v_m; k)}{dv_m}, \quad m = 1, \cdots, N. \tag{12}$$

2. $h(\cdot, \cdot)$ in (11) and $g(\cdot, \cdot)$ in (8) are related as:

$$g(v_m, v_j; k) = \frac{\partial h(v_m, v_j; k)}{\partial v_m} + \frac{\partial h(v_j, v_m; k)}{\partial v_m}. \tag{13}$$

Theorem 1 ((11) is a Lyapunov Function). *If (12) and (13) hold and if the coupling is symmetric (10), then the function $L(\cdots)$ defined in (11) is a Lyapunov function for the system (7).*

Proof. First, note that

$$\frac{dL}{dt} = \sum_{m=1}^{N} \frac{\partial L}{\partial v_m} \frac{dv_m}{dt}. \tag{14}$$

Expand

$$\frac{\partial L}{\partial v_m} = -\left[\frac{dz(v_m;\,k)}{dv_m} - \frac{\partial}{\partial v_m}\left(\sum_{i,j=1}^{N} J_{ij}\,h(v_i,v_j;\,k)\right)\right]$$

$$= -\left[\frac{dz(v_m;\,k)}{dv_m} - \sum_{i,j=1}^{N} J_{ij}\left(\frac{\partial h(v_i,v_j;\,k)}{\partial v_i}\delta_{im} + \frac{\partial h(v_i,v_j;\,k)}{\partial v_j}\delta_{jm}\right)\right]$$

$$= -\left[\frac{dz(v_m;\,k)}{dv_m} - \sum_{i,j=1}^{N} J_{ij}\frac{\partial h(v_i,v_j;\,k)}{\partial v_i}\delta_{im} - \sum_{i,j=1}^{N} J_{ij}\frac{\partial h(v_i,v_j;\,k)}{\partial v_j}\delta_{jm}\right]$$

$$= -\left[\frac{dz(v_m;\,k)}{dv_m} - \sum_{j=1}^{N} J_{mj}\frac{\partial h(v_m,v_j;\,k)}{\partial v_m} - \sum_{i=1}^{N} J_{im}\frac{\partial h(v_i,v_m;\,k)}{\partial v_m})\right]$$

$$= -\left[\frac{dz(v_m;\,k)}{dv_m} - \sum_{j=1}^{N} J_{mj}\frac{\partial h(v_m,v_j;\,k)}{\partial v_m} - \sum_{j=1}^{N} J_{jm}\frac{\partial h(v_j,v_m;\,k)}{\partial v_m}\right]$$

$$= -\left[\frac{dz(v_m;\,k)}{dv_m} - \sum_{j=1}^{N} J_{mj}\frac{\partial h(v_m,v_j;\,k)}{\partial v_m} - \sum_{j=1}^{N} J_{mj}\frac{\partial h(v_j,v_m;\,k)}{\partial v_m}\right] \quad (using\ (10))$$

$$= -\left[\frac{dz(v_m;\,k)}{dv_m} - \sum_{j=1}^{N} J_{mj}\left(\frac{\partial h(v_m,v_j;\,k)}{\partial v_m} + \frac{\partial h(v_j,v_m;\,k)}{\partial v_m}\right)\right] \quad (15)$$

$$= -\left[f(v_m;\,k) - \sum_{j=1}^{N} J_{mj}\,g(v_m,v_j;\,k)\right] \quad (using\ (12)\ and\ (13))$$

$$= -\frac{dv_m}{dt} \quad (using\ (7)). \quad (16)$$

Using (16), (14) becomes

$$\frac{dL}{dt} = -\sum_{m=1}^{N}\left(\frac{dv_m}{dt}\right)^2 \le 0, \quad (17)$$

proving that $L(\cdots)$ is non-increasing in t, hence constitutes a Lyapunov function for (7).   ∎

Lemma 1 (Stable equilibria and Lyapunov local minima are identical).
Any stable equilibrium of the generalized system (7) is a local minimum of the generalized Lyapunov function (11); and vice versa.

Proof. *Using (16), we can write (7) as*

$$\frac{dv_i}{dt} = -\frac{\partial L}{\partial v_i}, \quad i = 1,\cdots,N. \quad (18)$$

Given any equilibrium point $\vec{v}^* \triangleq [v_1^*,\cdots,v_N^*]^T$, i.e., $\frac{dv_i}{dt} = 0$, $\forall i$. By (18), the partial derivatives $\frac{\partial L(\vec{v}^*)}{\partial v_i} = 0$, $\forall i$, i.e., the equilibrium point is a local extremum (maximum/minimum/saddle/etc.. point) of $L(\cdots)$.

*Suppose $\vec{v}^*$ is a **stable** equilibrium. This means that there exists some ball $\mathcal{B}$ around $\vec{v}^*$, with radius greater than zero, such that if the system is perturbed to any point $\vec{v}^* + \delta\vec{v} \in \mathcal{B}$, the system's dynamics will return it to $\vec{v}^*$. More precisely, the projection of the derivative $\frac{d(\vec{v}^*+\delta\vec{v}(t))}{dt}$ onto the perturbation $\delta\vec{v}$ should be negative, i.e.,*

$$\delta\vec{v}^T \frac{d}{dt}(\vec{v}^* + \delta\vec{v}) < 0, \quad \forall\, \delta\vec{v} \text{ such that } \vec{v}^* + \delta\vec{v} \in \mathcal{B}. \tag{19}$$

Using (18), (19) becomes

$$\underbrace{\left.\frac{\partial L}{\partial \vec{v}}\right|_{\vec{v}^*+\delta\vec{v}}}_{\text{Jacobian of } L \text{ w.r.t } \vec{v} \text{ (row vector)}} \delta\vec{v} > 0, \quad \forall\, \delta\vec{v} \text{ such that } \vec{v}^* + \delta\vec{v} \in \mathcal{B}. \tag{20}$$

Since $L(\vec{v}; k)$ is differentiable (Definition 1), we have

$$L(\vec{v}^*) \simeq L(\vec{v}^* + \delta\vec{v}) - \left.\frac{\partial L}{\partial \vec{v}}\right|_{\vec{v}^*+\delta\vec{v}} \delta\vec{v} \Leftrightarrow L(\vec{v}^* + \delta\vec{v}) - L(\vec{v}^*) \simeq \left.\frac{\partial L}{\partial \vec{v}}\right|_{\vec{v}^*+\delta\vec{v}} \delta\vec{v}, \tag{21}$$

*with equality as $\delta\vec{v} \to \vec{0}$. Using (20) in (21), we have $L(\vec{v}^* + \delta\vec{v}) - L(\vec{v}^*) > 0$ for all $\delta\vec{v}$ such that $\vec{v}^* + \delta\vec{v}$ is in some ball $\mathcal{B}_2 \subset \mathcal{B}$, proving that v^* is a local **minimum** of $L(\vec{v}; k)$.*

Moreover, every step of the above argument can be reversed, proving that any local minimum of $L(\vec{v}; k)$ is a stable equilibrium point of (7). ∎

3.2 Bistability Properties of the Generalized System; Lyapunov–Ising-Hamiltonian Relation

First, we recall the (discrete) Ising Hamiltonian and establish a basic property.

Definition 2 (Discrete Ising Hamiltonian). *Given N "spins" (binary variables with values ± 1) $\{s_i\}$ and a set of coupling weights J_{ij} obeying (10), the (discrete) Ising Hamiltonian of the system is*

$$H(s_1, \cdots, s_N) \triangleq -\frac{1}{2}\sum_{i,j=1}^{N} J_{ij}\, s_i s_j. \tag{22}$$

Denoting $\vec{s} = [s_1, \cdots, s_N]^T$, this can also be written as $H(\vec{s})$.

Lemma 2 (Scaled/shifted Ising Hamiltonians preserve total order). *For any $k_1 > 0$ and any k_2, define a scaled/shifted version of the Ising Hamiltonian to be*

$$\tilde{H}(\vec{s}) \triangleq k_1 H(\vec{s}) + k_2. \tag{23}$$

Then, for any $\vec{s}_1, \vec{s}_2$ such that $H(\vec{s}_1) \leq H(\vec{s}_2)$, $\tilde{H}(\vec{s}_1) \leq \tilde{H}(\vec{s}_2)$; and vice-versa. This also implies that any local minimum of $H(\cdot)$ is a local minimum of $\tilde{H}(\cdot)$, and vice versa.

Proof. $\tilde{H}(\cdot)$ *is strictly monotonic and invertible with respect to* $H(\cdot)$, *establishing both directions of the first claim. The second claim follows from the first by contradiction.*

Next, we make an assumption about the bistability of $f(\cdot; k)$ when the gain k is high.

Assumption 3 (Bistability of each latch). $f(v; k)$ *is bistable if* $k = K$, *for some sufficiently large gain* $K > 0$; *i.e., for some* v_+, v_-, *with* $v_+ > v_-$,

$$f(v_+; K) = f(v_-; K) = 0. \tag{24}$$

Moreover,

$$\left.\frac{df(v; K)}{dv}\right|_{v=v_+} < 0 \quad and \quad \left.\frac{df(v; K)}{dv}\right|_{v=v_-} < 0. \tag{25}$$

These conditions ensure stable equilibria of $\frac{dv}{dt} = f(v; K)$ *(i.e., each equation of the generalized system (7), in the absence of coupling) at* v_+ *and* v_-. *Moreover, we assume that for each latch,* v_+ *and* v_- *are the only stable equilibria. This implies that* $v_i \in \{v_+, v_-\}$, $i = 1, ..., N$ *represent all the stable equilibria of (7) in the absence of coupling.*

We now make assumptions on the values of the functions $z(\cdot; \cdot)$ and $h(\cdot, \cdot; \cdot)$ at the bistable values v_+ and v_- when the gain is high. These assumptions are abstracted from properties of the $\tanh(\cdot)$ model (8).

Assumption 4 (Values for $z(\{v_+, v_-\}; K)$ **and** $h(\{v_+, v_-\}, \{v_+, v_-\}; K))$

$$z(v_+; K) = z(v_-; K) = c_3 \tag{26}$$
$$h(v_+, v_+; K) = h(v_-, v_-; K) = c_1, \tag{27}$$
$$h(v_+, v_-; K) = h(v_-, v_+; K) = c_2. \tag{28}$$

for some values c_1, $c_2 > c_1$ *and* c_3.

We can now establish a relation between the generalized Lyapunov function $L(\cdots)$ in (7) and the Ising Hamiltonian (22).

Theorem 2 (The Lyapunov function equals a scaled/shifted Ising Hamiltonian at nominal bistable values). *For* $i = 1, \cdots, N$, *if* $v_i \in \{v_+, v_-\}$, *define a corresponding "spin"* s_i *to be*

$$s_i = \begin{cases} 1 & if \ v_i = v_+, \\ -1 & if \ v_i = v_-. \end{cases} \tag{29}$$

Denote $\vec{s} \triangleq [s_1, \cdots, s_N]^T$ *and* $\vec{v}_B \triangleq [v_1, \cdots, v_N]^T$. *Then the generalized Lyapunov function* $L(\vec{v}_B; K)$ *equals a scaled/shifted version of the discrete Ising Hamiltonian* $H(\vec{s})$.

Proof. We have (using (26))

$$L(\vec{v}_B; K) = -\sum_{i=1}^{N}\left(z(v_i; K) - \sum_{j=1}^{N} J_{ij}\, h(v_i, v_j; K)\right)$$

$$= -Nc_3 + \sum_{i,j=1}^{N} J_{ij}\, h(v_i, v_j; K). \tag{30}$$

Defining

$$\tilde{h}(v_i, v_j) \triangleq \frac{2h(v_i, v_j; K) - (c_1 + c_2)}{c_2 - c_1} \tag{31}$$

$$\Leftrightarrow h(v_i, v_j; K) = \frac{(c_2 - c_1)\tilde{h}(v_i, v_j) + (c_1 + c_2)}{2},$$

(30) becomes

$$L(\vec{v}_B; K) = -Nc_3 + \frac{1}{2}\sum_{i,j=1}^{N} J_{ij}\left[(c_2 - c_1)\, \tilde{h}(v_i, v_j) + (c_1 + c_2)\right]$$

$$= -Nc_3 + \frac{c_1 + c_2}{2}\left(\sum_{i,j=1}^{N} J_{ij}\right) + \frac{c_2 - c_1}{2}\sum_{i,j=1}^{N} J_{ij}\, \tilde{h}(v_i, v_j). \tag{32}$$

From definition (31), note that

$$\tilde{h}(v_+, v_+) = \tilde{h}(v_-, v_-) = -1, \quad \tilde{h}(v_+, v_-) = \tilde{h}(v_-, v_+) = +1. \tag{33}$$

Hence, since $v_i \in [v_+, v_-]$, we have

$$\tilde{h}(v_i, v_j) = -s_i s_j. \tag{34}$$

Using (34) in (32), we have

$$L(\vec{v}_B; K) = -Nc_3 + \frac{c_1 + c_2}{2}\left(\sum_{i,j=1}^{N} J_{ij}\right) - \frac{c_2 - c_1}{2}\sum_{i,j=1}^{N} J_{ij}\, s_i s_j$$

$$= \underbrace{-Nc_3 + \frac{c_1 + c_2}{2}\left(\sum_{i,j=1}^{N} J_{ij}\right)}_{k_2} + \underbrace{(c_2 - c_1)}_{k_1 > 0} H(\vec{s}). \tag{35}$$

This is a scaled/shifted version of the Ising Hamiltonian (23). ∎

This immediately implies

Corollary 1 (Total order correspondence between Hamiltonian and Lyapunov functions). *For $i = 1, \cdots, N$, let*

$$\vec{v}_A \triangleq [v_{A,1}, \cdots, v_{A,N}]^T, \quad \vec{v}_B \triangleq [v_{B,1}, \cdots, v_{B,N}]^T,$$
$$\text{with } v_{A,i} \in \{v_+, v_-\},\ v_{B,i} \in \{v_+, v_-\}. \tag{36}$$

Let $\vec{s}_A$ and $\vec{s}_B$ be the spin vectors (defined using (29)) corresponding to $\vec{v}_A$ and $\vec{v}_B$, respectively. If $L(\vec{v}_A; K) \leq L(\vec{v}_B; K)$, then $H(\vec{s}_A) \leq H(\vec{s}_B)$; and vice versa.

Proof. Follows from Lemma 2 and Theorem 2.

As a result, any global minimum of one is also one of the other:

Corollary 2 (Hamiltonian and Lyapunov global minima correspond under bistability). *If $\vec{v}_A$ (36) is a global minimum of $L(\vec{v}; K)$ over all $\vec{v}$ with components taking bistable values v_+ or v_-, then the corresponding spin vector s_A (29) is a global minimum of $H(\vec{s})$; and vice versa.*

Proof. Follows from Corollary 1.

Even with coupling present in (7), we assume that each latch remains bistable, with only small deviations from v_+ and v_-:

Assumption 5 (Bistability persists in the presence of coupling). *In the presence of coupling, the exact values v_+ and v_- (Assumption 3) no longer represent stable equilibria for each latch, due to the perturbations introduced by the coupling. If the coupling is small enough, each latch will still have stable equilibria at some values v_{i+} and v_{i-} which are small perturbations of v_+ and v_-, respectively. This follows from the stability of the unperturbed equilibria. We assume, more generally, that this is true whether or not the coupling is small. More precisely, we assume that if $k = K$, then $v_i \in \{v_{i+}, v_{i-}\}$, with $v_{i+} \in [v_+ - \epsilon, v_+ + \epsilon]$ and $v_{i-} \in [v_- - \epsilon, v_- + \epsilon]$, for some $\epsilon \ll v_+ - v_-$, $\forall i = 1, \cdots, N$ capture **all stable equilibrium points** of (7).*

Assumption 5 enables us to benefit from Theorem 2 at the actual equilibrium points of (7):

Corollary 3 (Lyapunov function approximates a scaled/shifted Ising Hamiltonian at bistable values). *The Lyapunov function evaluated at bistable equilibrium points in the presence of coupling (as given in Assumption 5) approximates the scaled/shifted Ising Hamiltonian (35) at corresponding spin values.*

Proof. Follows from continuity of the Lyapunov function (11) in its arguments $v_1, \cdots, v_N$, and the fact that the bistable equilibrium points under coupling are small perturbations (Assumption 5) of the nominal bistable equilibria of Theorem 2.

Finally, note that the discrete Ising Hamiltonian (Definition 2) remains unchanged when all spins are flipped, since each term $s_i s_j$ does not change. Correspondingly, the Lyapunov function (11) remains unchanged if v_i is "flipped" from v_+ to v_-, and vice-versa (because of properties (26) to (28)). This implies that the search space of all Boolean combinations can be reduced by half; one spin can simply be set to either $+1$ or -1, and all combinations of the other spins explored. Correspondingly, for *one* chosen $i \in \{1, \cdots, N\}$, v_i can be set to either v_+ or v_-. We concretize this as

Corollary 4. *Let* $k = K$ *and* $v_i \in \{v_+, v_-\}$ *for* $i \in 1, \cdots, N$. *Then* v_N *can be fixed at* v_+ *without loss of generality, i.e., every value of the Lyapunov function* (11) $L(v_1, \cdots, v_N; K)$ *that can be achieved without this restriction can also be achieved with this restriction.*

4   Illustrative Examples

We now specialize the above results for our simple latch model of Sect. 2.2 and illustrate BLIM on fully-connected 3-spin graphs, as well as on G22, a 2000-spin, sparsely connected, MAX-CUT benchmark.

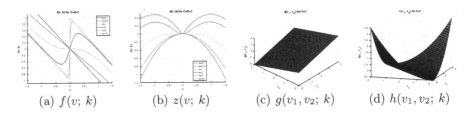

(a) $f(v; k)$ (b) $z(v; k)$ (c) $g(v_1, v_2; k)$ (d) $h(v_1, v_2; k)$

Fig. 4. Plots of $f()$, $g()$, $z()$ and $h()$ for $C = G = 1$.

First, we return to our $\tanh(\cdot)$ latch model of (6) and devise a specific Lyapunov function ((11)) for it. Recall that $f(\cdot; k)$ and $g(\cdot, \cdot; k)$ for this model are given in (8). Choosing the high value of gain (at which each latch features two stable states, see Assumption 3) to be

$$K \triangleq 5, \tag{37}$$

we solve $f(v; K) = 0$ numerically to obtain[8]

$$v_+ \triangleq 0.999909 \simeq +1, \quad v_- \triangleq -0.999909 = -v_+ \simeq -1,$$
$$\frac{df}{dv}(v_+) = -0.999999 < 0, \quad \frac{df}{dv}(v_-) = -0.999999 < 0. \tag{38}$$

Hence Assumption 3 is satisfied.[9] Assumption 5 can always be satisfied by making the couplings small enough. Define

$$z(v; k) \triangleq \int_0^v f(x; k)\, dx. \tag{39}$$

This obviously satisfies the requirement (12). Since $f(v; k)$ is odd in v (i.e., $f(-v; k) = -f(v; k)$, as is easily verified), it is easily shown that $z(v; k)$ in (39) is even, i.e.,

$$z(-v; k) = z(v; k). \tag{40}$$

---

[8] It is easy to show graphically that v_+, v_- and 0 are the only solutions of $f(v, K) = 0$.
[9] The third solution, $v = 0$, is unstable: $\frac{df}{dv}(0) = 24 > 0$.

Hence the requirement (26) is satisfied. Now define

$$h(v_i, v_j; k) \triangleq \frac{1}{2C} \left[\frac{(v_i - v_j)^2}{2} - 1 \right]. \tag{41}$$

It is easily verified that $h(v_i, v_j; k)$ satisfies the requirement (13). The additional requirements (27) and (28) are also satisfied, with

$$c_1 = -\frac{1}{2C}, \quad c_2 = \frac{1}{2C}(2v_+^2 - 1) \simeq -c_1 = \frac{1}{2C} > c_1. \tag{42}$$

Hence, using the definitions in (37) to (39) and (41) for K, v_+, v_-, $z(v; k)$ and $h(v_i, v_j; k)$, the coupled latch system (6) satisfies all the conditions needed for Theorem 1, Theorem 2, and their implications Lemma 1, Corollary 1, Corollary 2 and Corollary 3 to be valid. To summarize:

1. the behaviour of a system of coupled latches (6), for any latch gain k, is governed by a Lyapunov function which it minimizes locally to reach stable equilibria;
2. when the gain is high enough to make each latch bistable ($k = K$), a scaled/shifted version of the Lyapunov function closely approximates the system's discrete Ising Hamiltonian (Definition 2). This implies that if $H(\vec{s}_A) \leq H(\vec{s}_B)$ for any two spin states, the corresponding voltage states $\vec{v}_A$ and v_B obey $L(\vec{v}_A; K) \leq L(\vec{v}_A; K)$; and vice versa.[10] Moreover, global minima of the Ising Hamiltonian correspond to global minima of the Lyapunov function.

Plots of $f(v; k)$, $g(v_1, v_2; k)$, $z(v; k)$ and $h(v_1, v_2; k)$—equations (8), (39) evaluated numerically, and (41)—are shown in Fig. 4, for different values of the latch gain k.

4.1 Fully-Connected 3-Spin Graphs with Weights of Equal Magnitude

For insight, we explore BLIM on all fully-connected 3-spin graphs with weights of equal magnitude. We choose 3-spin graphs because their Lyapunov functions can be visualized completely in three dimensions.

A three-spin graph is a triangle, i.e., with three vertices and three edges with weights J_{12}, J_{23} and J_{13}. The Ising Hamiltonian (Definition 2) is

$$H_3(s_1, s_2, s_3) \triangleq -\frac{1}{2} \left(J_{12}\, s_1 s_2 + J_{13}\, s_2 s_3 + J_{23} s_1 s_3 \right), \tag{43}$$

and the Lyapunov function becomes

$$L_3(v_1, v_2, v_3; k) \triangleq -z(v_1; k) - z(v_2; k) - z(v_3; k) + 2\big[J_{12}\, h(v_1, v_2; k)$$
$$+ J_{23}\, h(v_2, v_3; k) + J_{13}\, h(v_1, v_3; k) \big], \tag{44}$$

---

[10] Recall that $H(\vec{\cdot})$ is the Ising Hamiltonian and $L(\vec{\cdot}; \cdot)$ the Lyapunov function.

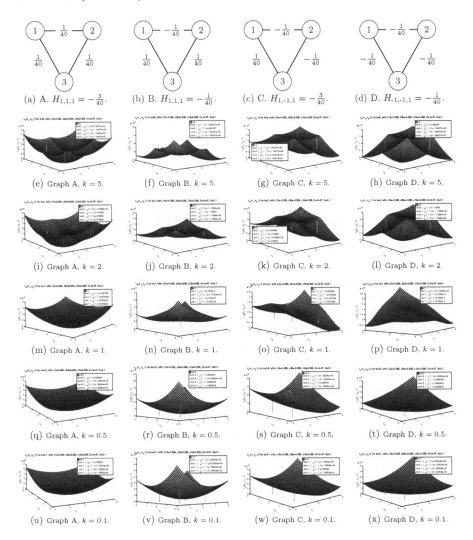

Fig. 5. $L_3(v_1, v_2, 1)$ for fully-connected 3-node graphs.

where $f(\cdot; k)$ and $h(\cdot, \cdot; k)$ are given by (39) and (41), and we have used the even symmetry of $h(\cdot, \cdot; k)$. Also, applying Corollary 4, we set $v_3 = v_+$, which turns (44) into

$$L_3(v_1, v_2, v_+; k) \triangleq -z(v_1; k) - z(v_2; k) - z(v_+; k) + 2\big[J_{12}\, h(v_1, v_2; k) \\ + J_{23}\, h(v_2, v_+; k) + J_{13}\, h(v_1, v_+; k)\big], \tag{45}$$

Consider fully-connected graphs with edge weights $\pm A_c$, where A_c is a coupling strength parameter. There are only 4 unique fully-connected 3-node graphs of

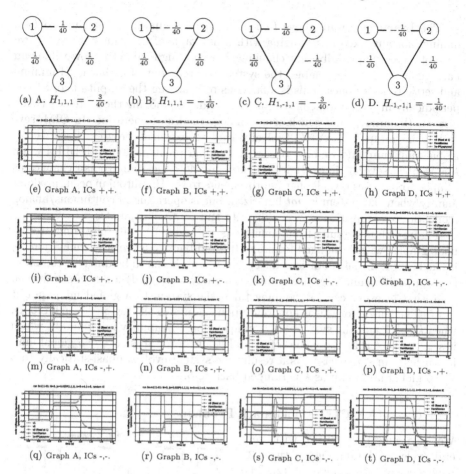

Fig. 6. Simulations of (7) and (4) with different initial conditions for fully-connected 3-node graphs.

this type, as shown in the top row[11] of Fig. 5—other fully-connected graph possibilities are congruent to one of these. Figure 5 depicts the Lyapunov functions for all four fully-connected graphs—each column shows the Lyapunov functions for various values of k for a particular graph. The points corresponding to discrete Ising spins, i.e., $v_1, v_2 = \pm 1$, are marked with vertical lines and the Lyapunov values at these points are noted in the legend. As expected, high values of k feature multiple local minima, which coalesce as k is lowered below 1. For the case of weights $(-\frac{1}{40}, -\frac{1}{40}, -\frac{1}{40})$, where several points reach the global minimum, the Lyapunov landscape for low k becomes more of a saddle region than a single well-defined global minimum.

---

[11] Each node represents an Ising spin; the weight of the edge between two nodes i and j is J_{ij}.

The Lyapunov result above (Theorem 1) only guarantees settling to *local* minima, for any fixed (unchanging with time) value of the gain k. However, we have observed empirically that changing k from a high value to a low value and back again (over time) enables the system to break out of higher local minima and settle to lower ones. This is analogous to changing the amplitude of SYNC periodically in OIM and achieves the same end, *i.e.*, moving the system between discrete (binarized) and continuous (analog) modes of operation. In OIM, several such "ramps" of SYNC typically lead to excellent progress towards the global minimum, and we have observed a similar phenomenon with BLIM when k is ramped several times. Indeed, examining how the Hamiltonian changes as ramping progresses reveals that improvements to the Hamiltonian occur predominantly when the system is *not* binarized, but is operating in continuous/analog mode, or is in transition between discrete and continuous modes.

Figure 6 shows results from simulating (7) and (8) for all fully-connected 3-node graphs, with k changed from 5 to 0.1 and back to 5 over the simulation. v_3 is fixed at $+1$; initial conditions for v_1 and v_2 are chosen to be different combinations of positive and negative values (randomly generated) for each simulation. Note that in every case, k ramping takes the system to a global minimum at $k = 5$. The reason for this is apparent upon examining the region $t \sim [60, 120]\mu s$, when $k = 0.1$ and the system has a unique global Lyapunov minimum to which it settles. Note that the values of v_1 and v_2 "lean towards" a $k = 5$ global minimum here, in every case. As a result, the system evolves quasi-statically to a $k = 5$ global minimum as k is ramped back to 5.

4.2   G22 MAX-CUT Benchmark Problem

We also illustrate BLIM with several cycles of k-ramping on the G22 Ising benchmark problem [1,12]. The problem has $N = 2000$ nodes/spins, sparsely interconnected (with randomly generated ± 1 weights) with 19,990 connections. Figure 7 shows the progress of the Ising Hamiltonian as BLIM runs on this problem, with k ramped from 0.5 to 2 in a square wave fashion about 6 times over the simulation. As can be seen, subsequent cycles of k-ramping reduce the Hamiltonian; moreover, the reduction hap-

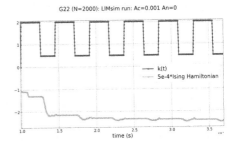

Fig. 7. BLIM on the 2000-spin Ising benchmark problem G22 [1,12]: the Hamiltonian improves when k is low, *i.e.*, the system is in "analog mode".

pens largely when k is low, *i.e.*, the system is *not* binarized, but in analog mode. As with OIM and other Ising machine schemes, the underlying mechanism behind this is not well understood at this point but it is this feature, that some deep analog mechanism is solving the originally discrete Ising problem, that fundamentally separates Ising machines from hardware or software implementations of essentially discrete minimization methods [9,11].

5   Conclusion

We have presented BLIM, an Ising machine scheme based on latches (bistable elements) with controllable gains. Using a simple dynamical model that distills the essence of a back-to-back inverter-based latch, we have set up equations for coupled latch systems and shown that they can be generalized to a form that also captures coupled oscillator networks when two functions are defined appropriately. We have proved that under appropriate conditions, this generalized form has a Lyapunov function which becomes essentially identical to the Ising Hamiltonian when the system is driven to binarized states, *e.g.*, by making latch gains high. This result implies that the system will settle naturally to local minima of the Lyapunov function. Furthermore, varying the gains periodically is seen to lead the system to lower minima. Our general formulation enables side-by-side comparison of OIM, BLIM, and possibly other Ising machine schemes, which may lead to progress in unravelling the mechanisms that underlie Ising machines' intriguing global minimization tendencies. BLIM retains an important practical feature of OIM, *i.e.*, that it can be implemented using miniaturisable CMOS electronics; however, implementations in other physical domains, such as (synthetic) biology, may also be of interest.

Acknowledgments. We thank Tianshi Wang, Nagendra Krishnapura, Yiannis Tsividis and Peter Kinget for discussions that motivated this work. Support from the US National Science Foundation is gratefully acknowledged.

References

1. The G-set benchmarks for MAX-CUT. http://grafo.etsii.urjc.es/optsicom/maxcut
2. Bian, Z., Chudak, F., Macready, W.G., Rose, G.: The Ising model: teaching an old problem new tricks. D-Wave Systems 2 (2010)
3. Wang, T., Roychowdhury, J.: OIM: oscillator-based Ising machines for solving combinatorial optimisation problems. In: McQuillan, I., Seki, S. (eds.) UCNC 2019. LNCS, vol. 11493, pp. 232–256. Springer, Cham (2019). https://doi.org/10.1007/978-3-030-19311-9_19 Preprint available at arXiv:1903.07163 [cs.ET] at arXiv:1903.07163 [cs.ET]
4. Brush, S.G.: History of the Lenz-Ising model. Rev. Mod. Phys. **39**(4), 883–893 (1967). https://doi.org/10.1103/RevModPhys.39.883
5. Comorosan, S.: New mechanism for the control of cellular reactions: the biochemical flip-flop. Nature **227**(5253), 64–65 (1970)
6. Endo, K., Hayashi, K., Inoue, T., Saito, H.: A versatile cis-acting inverter module for synthetic translational switches. Nature Commun. **4** (2013). https://doi.org/10.1038/ncomms3393
7. Gärtner, B., Matousek, J.: Approximation Algorithms and Semidefinite Programming. Springer, Heidelberg (2014). https://doi.org/10.1007/978-3-642-22015-9
8. Johnson, M.W., et al.: Quantum annealing with manufactured spins. Nature **473**(7346), 194–198 (2011)
9. Camsari, K.Y., Faria, R., Sutton, B.M., Datta, S.: Stochastic p-bits for invertible logic. Phys. Rev. X **7**(3), 031014 (2017)

10. Kirkpatrick, S., Gelatt, C.D., Vecchi, M.P.: Optimization by simulated annealing. Science **220**(4598), 671–680 (1983)

11. Yamaoka, M., Yoshimura, C., Hayashi, M., Okuyama, T., Aoki, H., Mizuno, H.: A 20k-spin Ising chip to solve combinatorial optimization problems with CMOS annealing. IEEE J. Solid-State Circ. **51**(1), 303–309 (2016)

12. Festa, P., Pardalos, P.M., Resende, M.G.C., Ribeiro, C.C.: Randomized heuristics for the MAX-CUT problem. Opt. Methods Softw. **17**(6), 1033–1058 (2002)

13. Shichman, H., Hodges, D.A.: Modeling and simulation of insulated-gate field-effect transistor switching circuits. IEEE J. Solid-State Ckts. **3**(3), 285–289 (1968)

14. Inagaki, T., et al.: A Coherent Ising machine for 2000-node optimization problems. Science **354**(6312), 603–606 (2016)

15. Wang, T., Roychowdhury, J.: Oscillator-based Ising machine. arXiv:1709.08102 (2017)

16. Wang, T., Wu, L., Roychowdhury, J.: New computational results and hardware prototypes for oscillator-based Ising machines. In: Proceedings of the IEEE DAC, pp. 239:1–239:2 (2019). https://doi.org/10.1145/3316781.3322473

17. Wang, T., Wu, L., Nobel, P., Roychowdhury, J.: Solving combinatorial optimisation problems using oscillator based Ising machines. Natural Comput., 1–20 (2021)

18. Haribara, Y., Utsunomiya, S., Yamamoto, Y.: Computational principle and performance evaluation of Coherent Ising Machine based on degenerate optical parametric oscillator network. Entropy **18**(4), 151 (2016)

19. Wang, Z., Marandi, A., Wen, K., Byer, R.L., Yamamoto, Y.: Coherent Ising machine based on degenerate optical parametric oscillators. Phys. Rev. A **88**(6), 063853 (2013)

Physical ZKP for Connected Spanning Subgraph: Applications to Bridges Puzzle and Other Problems

Suthee Ruangwises[✉][iD] and Toshiya Itoh[iD]

Department of Mathematical and Computing Science,
Tokyo Institute of Technology, Tokyo, Japan
titoh@c.titech.ac.jp

Abstract. An undirected graph G is known to both the prover P and the verifier V, but only P knows a subgraph H of G. Without revealing any information about H, P wants to convince V that H is a connected spanning subgraph of G, i.e. H is connected and contains all vertices of G. In this paper, we propose an unconventional zero-knowledge proof protocol using a physical deck of cards, which enables P to physically show that H satisfies the condition without revealing it. We also show applications of this protocol to verify solutions of three well-known NP-complete problems: the Hamiltonian cycle problem, the maximum leaf spanning tree problem, and a popular logic puzzle called Bridges.

Keywords: Zero-knowledge proof · Card-based cryptography · Connected spanning subgraph · Hamiltonian cycle · Maximum leaf spanning tree · Graph · Bridges · Puzzle

1 Introduction

A *zero-knowledge proof (ZKP)* is an interactive protocol introduced by Goldwasser et al. [8], which enables a prover P to convince a verifier V that a statement is correct without revealing any other information. A ZKP with perfect completeness and soundness must satisfy the following three properties.

1. **Perfect Completeness:** If the statement is correct, then V always accepts.
2. **Perfect Soundness:** If the statement is incorrect, then V always rejects.
3. **Zero-knowledge:** During the verification, V gets no extra information other than the correctness of the statement. Formally, there exists a probabilistic polynomial time algorithm S (called a *simulator*), without an access to P but with a black-box access to V, such that the outputs of S follow the same probability distribution as the outputs of the actual protocol.

Goldreich et al. [7] proved that a computational ZKP exists for every NP problem. Several recent results, however, instead considered an unconventional way of constructing ZKPs by using physical objects such as a deck of cards and

© Springer Nature Switzerland AG 2021
I. Kostitsyna and P. Orponen (Eds.): UCNC 2021, LNCS 12984, pp. 149–163, 2021.
https://doi.org/10.1007/978-3-030-87993-8_10

envelopes. The benefit of these physical protocols is that they allow external observers to check that the prover truthfully executes the protocol (which is often a challenging task for digital protocols). They also have didactic values and can be used to teach the concept of ZKP to non-experts.

Consider a verification of the following condition. An undirected graph G is known to both P and V, but only P knows a subgraph H of G. Without revealing any information about H, P wants to convince V that H is a connected spanning subgraph of G, i.e. H is connected and contains all vertices of G.

A ZKP to verify the connected spanning subgraph condition is important because this condition is a part of many well-known NP-complete problems, such as the Hamiltonian cycle problem, the maximum leaf spanning tree problem, and a famous logic puzzle called *Bridges*. To verify solutions of these problems, P needs to show that his/her solution satisfies the connected spanning subgraph condition as well as some other conditions (which are relatively easier to show).

1.1   Related Work

Most of previous work in physical ZKPs aimed to verify a solution of popular logic puzzles: Sudoku [9,20], Nonogram [4], Akari [2], Kakuro [2,14], KenKen [2], Takuzu [2,13], Makaro [3], Norinori [5], Slitherlink [12], Juosan [13], Numberlink [18], Suguru [17], Ripple Effect [19], Nurikabe [16], and Hitori [16].

The theoretical contribution of these protocols is that they employ novel methods to physically verify specific functions. For example, a subprotocol in [3] verifies that a number in a list is the largest one in that list without revealing any value in the list, and a subprotocol in [9] verifies that a list is a permutation of all given numbers without revealing their order.

Some of these protocols can verify graph theoretic problems. For example, a protocol in [18] verifies a solution of the k vertex-disjoint paths problem, i.e. a set of k vertex-disjoints paths joining each of the k given pairs of endpoints in a graph. In a recent work, a subprotocol in [16] also verifies a condition related to connectivity. However, their protocol only works in a grid graph and also deals with a different condition from the one considered in this paper. (Their protocol only verifies that the selected cells on a board are connected together, not as a spanning subgraph of the whole board.)

1.2   Our Contribution

In this paper, we propose a physical card-based ZKP with perfect completeness and soundness to verify that a subgraph H is a connected spanning subgraph of an undirected graph G without revealing H.

We also show three possible applications of this protocol: verifying a Hamiltonian cycle in an undirected graph, verifying the existence of a spanning tree with at least k leaves in an undirected graph, and verifying a solution of the Bridges puzzle.

2   Preliminaries

Each *encoding card* used in our protocol has either ♣ or ♡ on the front side. All cards have indistinguishable back sides.

For $0 \leq x < k$, define $E_k(x)$ to be a sequence of consecutive k cards, with all of them being ♣ except the $(x+1)$-th card from the left being ♡, e.g. $E_3(0)$ is ♡♣♣ and $E_4(2)$ is ♣♣♡♣. We use $E_k(x)$ to encode an integer x in $\mathbb{Z}/k\mathbb{Z}$. This encoding rule was introduced by Shinagawa et al. [22].

The cards in $E_k(x)$ are arranged horizontally as defined above unless stated otherwise. In some situations, however, we may arrange the cards vertically, where the leftmost card becomes the topmost card and the rightmost card becomes the bottommost card.

In an $m \times k$ *matrix* of cards, let Row i denote the i-th topmost row and Column j denote the j-th leftmost column.

2.1   Pile-Shifting Shuffle

A *pile-shifting shuffle* on an $m \times k$ matrix shifts the columns of the matrix by a random cyclic shift, i.e. shifts the columns cyclically to the right by r columns for a uniformly random $r \in \mathbb{Z}/k\mathbb{Z}$ unknown to all parties.

This protocol was developed by Shinagawa et al. [22]. It can be implemented in real world by putting the cards in each column into an envelope and applying several *Hindu cuts* to the sequence of envelopes [23].

2.2   Sequence Selection Protocol

Suppose we have k sequences $A_0, A_1, ..., A_{k-1}$, each encoding an integer in $\mathbb{Z}/m\mathbb{Z}$, and a sequence B encoding an integer b in $\mathbb{Z}/k\mathbb{Z}$. We propose the following *sequence selection protocol*, which allows us to select a sequence A_b (to be used as an input in other protocols) without revealing b.

1. Construct the following $(m+2) \times k$ matrix M (see Fig. 1).
   (a) In Row 1, place a sequence $E_k(0)$. In Row 2, place the sequence B.
   (b) In each Column $j = 1, 2, ..., k$, place the sequence A_{j-1} arranged vertically from Row 3 to Row $m+2$.
2. Apply the pile-shifting shuffle to M.
3. Turn over all cards in Row 2. Locate the position of a ♡. Suppose it is at Column j.
4. Select the sequence in Column j arranged vertically from Row 3 to Row $m+2$. This is the sequence A_b as desired. Turn over all face-up cards.

After we are done using A_b in other protocols, we can put A_b back into M, apply the pile-shifting shuffle to M, then turn over all cards in Row 1 and shift the columns of M cyclically such that the ♡ in Row 1 moves to Column 1. This reverts the matrix back to its original position, so we can reuse the sequences $A_0, A_1, ..., A_{k-1}$, and B.

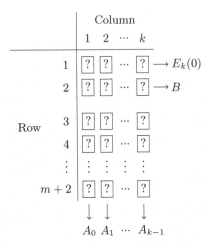

Fig. 1. An $(m+2) \times k$ matrix M constructed in Step 1

2.3   Enhanced Matrix

In addition to the encoding cards, we also use *marking cards*, each having a positive integer on the front side. All cards have indistinguishable back sides.

Starting from an $m \times k$ matrix of face-down encoding cards, place face-down marking cards $\boxed{1}$, $\boxed{2}$, ..., $\boxed{k}$ from left to right on top of Row 1; this new row is called Row 0. Then, place face-down marking cards $\boxed{2}$, $\boxed{3}$, ..., $\boxed{m}$ from top to bottom (starting at Row 2) to the left of Column 1; this new column is called Column 0. We call this structure an $m \times k$ enhanced matrix (see Fig. 2).

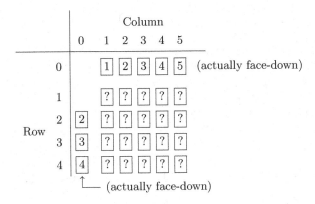

Fig. 2. An example of a 4×5 enhanced matrix

2.4   Double-Scramble Shuffle

In a *double-scramble shuffle* on an $m \times k$ enhanced matrix, first rearrange Columns $1, 2, ..., k$ (including the marking cards in Row 0) by a uniformly random permutation unknown to all parties (which can be implemented by putting the cards in each column into an envelope and scrambling all envelopes together). Then, leave Row 1 as it is and rearrange Rows $2, 3, ..., m$ (including the marking cards in Column 0) by a uniformly random permutation unknown to all parties. This protocol was developed by Ruangwises and Itoh [18].

2.5   Rearrangement Protocol

A *rearrangement protocol* reverts the rows and columns of an enhanced matrix (after we perform double scramble shuffles) back to their original positions so that we can reuse the cards without revealing them. This protocol was developed by Ruangwises and Itoh [18], although slightly different protocols with the same idea were also used in other previous work [3, 10, 11, 19, 20].

In the rearrangement protocol on an $m \times k$ enhanced matrix, first apply the double-scramble shuffle to the matrix. Then, turn over all marking cards in Row 0 and rearrange the columns such that each marking card with number i will be in Column i. Analogously, turn over all marking cards in Column 0 and rearrange Rows $2, 3, ..., m$ accordingly.

2.6   Neighbor Counting Protocol

Suppose we have an $m \times k$ matrix with each row encoding an integer in $\mathbb{Z}/k\mathbb{Z}$. A *neighbor counting protocol* allows us to count the number of indices $i \geq 2$ such that Row i encodes the same integer as Row 1, without revealing any other information. This protocol was developed by Ruangwises and Itoh [18].

1. Place marking cards to make the matrix become an $m \times k$ enhanced matrix.
2. Apply the double-scramble shuffle.
3. Turn over all encoding cards in Row 1. Locate the position of a ♡. Suppose it is at Column j.
4. Turn over all encoding cards in Column j. Count the number of ♡s besides the one in Row 1. This is the number of indices that we want to know.
5. Turn over all face-up cards. Apply the rearrangement protocol.

3   Verifying an Undirected Path

In this section, we will explain a *path verification protocol*, which verifies the existence of an undirected path between vertices s and t in an undirected graph

G. It is a special case $k = 1$ of the protocol for the k vertex-disjoint paths problem developed by Ruangwises and Itoh [18][1].

We call s and t *terminal vertices*, and other vertices *non-terminal vertices*. We call a path $(v_1, v_2, ..., v_\ell)$ *minimal* if there are no neighboring vertices v_i and v_j such that $|i - j| > 1$. Observe that given any path between s and t, one can modify it to become a minimal one in linear time, so we can assume that P knows a minimal path between s and t.

Let d be the maximum degree of a vertex in G. In linear time, we can color the vertices of G with at most $d + 1$ colors such that there are no neighboring vertices with the same color. This $(d + 1)$-coloring is known to all parties.

On each terminal vertex v, P publicly places a sequence $E_{d+2}(0)$. On each non-terminal vertex v with the x-th color, P secretly places a sequence $E_{d+2}(0)$ if v is on P's path, or a sequence $E_{d+2}(x)$ if v is not on the path. Let $A(v)$ denote the sequence on each vertex v. Since the path is minimal, every non-terminal cell on the path has exactly two neighbors with a sequence encoding the same number as it (which is 0), while every terminal cell has exactly one such neighbor. On the other hand, every non-terminal cell not on the path has no neighbor with a sequence encoding the same number as it.

The idea is that, for every vertex v with the x-th color, P will add two "artificial neighbors" of v, both having $E_{d+2}(x)$ on it, and show that

1. every non-terminal vertex v (both on and not on the path) has exactly two neighbors with a sequence encoding the same number as $A(v)$, and
2. every terminal vertex v has exactly one neighbor with a sequence encoding the same number as $A(v)$.

Formally, to verify each non-terminal (resp. terminal) vertex v with the x-th color and with degree d_v, P performs the following steps.

1. Construct the following $(d_v + 3) \times (d + 2)$ matrix M.
   (a) In Row 1, place $A(v)$.
   (b) In each of the next d_v rows, place $A(v')$ for each neighbor v' of v.
   (c) In each of the last two rows, place $E_{d+2}(x)$.
2. Apply the neighbor counting protocol to M. V verifies that there are exactly two rows (resp. one row) encoding the same integer as Row 1.
3. Put the sequences back to their corresponding vertices.

If every vertex in G passes the verification, then V accepts.

4   Verifying a Connected Spanning Subgraph

We get back to our main problem. Let $v_1, v_2, ..., v_n$ be the vertices in G. In order to prove that H is a connected spanning subgraph of G, it is sufficient

---

[1] Although the k vertex-disjoint paths problem is NP-complete when k is a part of the input, the special case $k = 1$ in solvable in linear time. Hence, this protocol is actually unnecessary since V can easily verifies existence of the path by him/herself given G. However, we explain the details of this protocol in order to show its idea, which will be modified and used in our main protocol in Sect. 4.

to show that there is an undirected path between v_i and v_n in H for every $i = 1, 2, ..., n - 1$.

Note that the path verification protocol in Sect. 3 verifies a path between s and t in a graph G, where G is known to all parties. In this section, we will modify that protocol so that it can verify a path between s and t in a subgraph H of G, where H is known to only P. Then, P will perform the modified protocol for $n - 1$ rounds, with $s = v_i$ and $t = v_n$ in each i-th round.

At the beginning, P secretly places a sequence $B(e)$ on every edge $e \in G$ to indicate whether $e \in H$. ($B(e)$ is $E_2(1)$ if $e \in H$ and is $E_2(0)$ if $e \notin H$.) By doing this, the graph H is committed and cannot be changed later.

Let d be the maximum degree of a vertex in G. Like in the path verification protocol, consider a $(d + 1)$-coloring, known to all parties, such that there are no neighboring vertices with the same color.

On every vertex v, P publicly places a sequence $A_0(v)$, which is $E_{d+3}(d+2)$. $A_0(v)$ acts as a "blank sequence" guaranteed to be different from $A_1(v')$ on any vertex v' during any round, which will be defined in the next step.

During each i-th round when P wants to show that there is a path in H between $s = v_i$ and $t = v_n$. First, P selects a minimal path between s and t in H. On each terminal vertex v, P publicly places a sequence $A_1(v)$, which is $E_{d+3}(0)$. On each non-terminal vertex v with the x-th color, P secretly places a sequence $A_1(v)$, which is $E_{d+3}(0)$ if v is on the path and is $E_{d+3}(x)$ if v is not on the path. Note that unlike $A_0(v)$ which remains the same throughout the whole protocol, $A_1(v)$ is changed in every round since it depends on the path selected in each round.

The verification steps are similar to the path verification protocol, except that in Step 1(b), P first applies the sequence selection protocol in Sect. 2.2 to determine whether to choose $A_0(v')$ or $A_1(v')$ for each neighbor v' of v, depending on whether an edge e between v and v' is in H or not. The idea is that if $e \in H$, then v' is still v's neighbor in H, so P chooses a sequence $A_1(v')$ and the rest works the same way as in the path verification protocol. On the other hand, if $e \notin H$, then v' is not v's neighbor in H, so P chooses a sequence $A_0(v')$ which is guaranteed to be different from $A_1(v)$.

Formally, to verify each non-terminal (resp. terminal) vertex v with the x-th color and with degree d_v, P performs the following steps.

1. Construct the following $(d_v + 3) \times (d + 3)$ matrix M.
   (a) In Row 1, place $A_1(v)$.
   (b) For each neighbor v' of v, let e be an edge between v and v', and let b be a bit encoded by $B(e)$. Apply the sequence selection protocol to choose a sequence $A_b(v')$ and place it in the next row of M. Repeatedly perform this for every neighbor of v to fill the next d_v rows.
   (c) In each of the last two rows of M, place $E_{d+3}(x)$.
2. Apply the neighbor counting protocol to M. V verifies that there are exactly two rows (resp. one row) encoding the same integer as Row 1.
3. Put the sequences back to their corresponding vertices.

If every vertex in G passes the verification, then V accepts.

This protocol uses $2(d + 3)(2n + 2) + 2d + 2m$ encoding cards and $2d + 5$ marking cards, where n and m are the numbers of vertices and edges of G, respectively, and d is the maximum degree of a vertex in G. Therefore, the total number of required cards is $\Theta(dn)$.

5   Proof of Correctness and Security

We will prove the perfect completeness, perfect soundness, and zero-knowledge properties of our main protocol in Sect. 4.

Lemma 1 (Perfect completeness). *If H is a connected spanning subgraph of G, then V always accepts.*

Proof. Suppose that H is a connected spanning subgraph of G, then there exists a path between v_i and v_n in H for every $i = 1, 2, ..., n - 1$.

First, we will prove the correctness of the sequence selection protocol in Sect. 2.2. Since B encodes the number b, when placing B in Row 2, the $\boxed{\heartsuit}$ will be at Column $b + 1$, the same column as the sequence A_b. After applying the pile-shifting shuffle, they will still be at the same column, so the sequence we get in Step 4 will be A_b.

Now consider the main protocol in each i-th round. In Step 1(b), P always selects a sequence $A_1(v')$ if $e \in H$ and $A_0(v')$ if $e \notin H$. Since $A_0(v')$ is $E_{d+3}(d+2)$ and thus is different from $A_1(v)$, adding $A_0(v')$ to a new row of M does not increase the number of rows encoding the same integer as Row 1. Therefore, the result will remain the same even if in Step 1(b) P adds only the sequences on the vertices such that $e \in H$, which is equivalent to solely applying the path verification protocol in Sect. 3 to verify a path between v_i and v_n on H.

The perfect completeness property of the path verification protocol has been proved in [18], so we can conclude that V always accepts.   □

Lemma 2 (Perfect soundness). *If H is not a connected spanning subgraph of G, then V always rejects.*

Proof. Suppose that H is not a connected spanning subgraph of G, then there exists an index $i \in \{1, 2, ..., n - 1\}$ such that there is no path between v_i and v_n in H. In Lemma 1, we have proved that the sequence selection protocol is correct, and the i-th round of the main protocol is equivalent to applying the path verification protocol to verify a path between v_i and v_n on H.

The perfect soundness property of the path verification protocol has been proved in [18], so we can conclude that V always rejects.   □

Lemma 3 (Zero-knowledge). *During the verification, V learns nothing about H.*

Proof. To prove the zero-knowledge property, it is sufficient to prove that all distributions of the values that appear when the cards are turned face-up can be simulated by a simulator S without knowing H.

- In the sequence selection protocol:
  - In Step 3, we turn over all cards in Row 2. This occurs right after a pile-shifting shuffle is applied to M. Hence, the ♡ has an equal probability to be at each of the k columns, so this step can be simulated by S.
- In the neighbor counting protocol:
  - In Step 3, we turn over all encoding cards in Row 1. The order of Columns $1, 2, ..., k$ is uniformly distributed among all possible permutations due to the double-scramble shuffle. Hence, the ♡ has an equal probability to be at each of the k columns, so this step can be simulated by S.
  - In Step 4, we turn over all encoding cards in Column j. Suppose there are t ♡s besides the one in Row 1 (t is now a public information). The order of Rows $2, 3, ..., m$ is uniformly distributed among all possible permutations due to the double-scramble shuffle. Hence, all t ♡s have an equal probability to be at each of the $\binom{m-1}{t}$ combinations of rows, so this step can be simulated by S.

Therefore, we can conclude that V learns nothing about H.   □

6 Applications to NP-Complete Problems

6.1 Hamiltonian Cycle Problem

Given an undirected graph G, determining whether G has a Hamiltonian cycle (a cycle that visits each vertex exactly once) is known to be NP-complete [6]. Suppose P knows a Hamiltonian cycle H of G and wants to convince V that G has a Hamiltonian cycle without revealing any information about H.

To prove that H is a Hamiltonian cycle of G, it is sufficient to show that

1. H is a connected spanning subgraph of G, and
2. every vertex in H has degree 2.

At the beginning, P commits H by secretly placing a sequence $B(e)$ on every edge $e \in G$ to indicate whether $e \in H$. ($B(e)$ is $E_2(1)$ if $e \in H$ and is $E_2(0)$ if $e \notin H$.) The first condition can be verified by the protocol in Sect. 4.

To verify the second condition, P first applies the copy protocol explained in Appendix A.1 to make another copy of a sequence $B(e)$ on every edge e. (Each of the two copies will be used to verify each endpoint of e.) For each vertex $v \in H$, P considers one (unused) copy of a sequence on every edge e incident to v and selects only the leftmost card of it (which is ♣ if $e \in H$ and is ♡ if $e \notin H$). Then, P scrambles all selected cards together and turns over all of them, and V verifies that there are exactly two ♣s among them (which means v has degree 2 in H). V accepts if the verification passes for every vertex in H. This protocol also uses $\Theta(dn)$ cards.[2]

---

[2] There is an alternative way to verify a Hamiltonian cycle: P publicly constructs an $n \times n$ adjacency matrix M of G, then privately selects a permutation σ and rearranges both the rows and columns of M by σ. Finally, P turns over all cards in the form $M(i, i+1)$ and $M(i, i-1)$ to show that they are all 1s. This protocol is simpler and more straightforward, but it requires $\Theta(n^2)$ cards, which is significantly greater than our protocol in sparse graphs.

6.2   Maximum Leaf Spanning Tree Problem

Given an undirected graph G and an integer k, the decision version of the maximum leaf spanning tree problem asks whether G has a spanning tree with at least k leaves (vertices with degree 1). This problem is also known to be NP-complete [6]. Suppose P knows a spanning tree H of G with at least k leaves and wants to convince V that the such tree exists without revealing any information about H.

To prove that G has a spanning tree with at least k leaves, it is sufficient to show that

1. H is a connected spanning subgraph of G, and
2. H has at least k leaves.

Note that it is not necessary to show that H itself is a tree. (Even if H itself is not a tree, any spanning tree of H will also be a spanning tree of G, and every leaf of H will still be a leaf of that tree, so G must have a spanning tree with at least k leaves.)

P commits H by the same way as in the Hamiltonian cycle problem, and uses the protocol in Sect. 4 to verify the first condition.

To verify the second condition, P makes an additional copy of every $B(e)$ like in the Hamiltonian cycle problem. For every vertex v, P selects only the leftmost card of $B(e)$ on every edge e incident to it, scramble these cards, and puts them into an envelope. (If there are less than d cards, P publicly adds more ♡s until there are d cards before scrambling them.) Then, P scrambles all envelopes together. Next, P picks an envelope, opens it and looks at the front side of all cards inside (without V seeing the front side). If there is exactly one ♣ among them, P reveals all cards to let V verify that there is exactly one ♣ (which means the corresponding vertex is a leaf); otherwise, P does not reveal the cards. P repeatedly does this for every envelope. V accepts if there are at least k envelopes with exactly one ♣. This protocol also uses $\Theta(dn)$ cards.

6.3   Bridges Puzzle

Bridges, or the Japanese name Hashiwokakero, is a logic puzzle created by a Japanese company Nikoli, which also developed many other popular logic puzzles including Sudoku, Kakuro, and Numberlink.

A Bridges puzzle consists of a rectangular grid of size $p \times q$, with some cells called *islands* containing an encircled positive number of at most 8. The objective of this puzzle is to connect some pairs of islands by straight lines called *bridges* that can only run horizontally or vertically. There can be at most two bridges between each pair of islands, and the bridges must satisfy the following conditions [15] (see Fig. 3).

1. *Island condition*: The number of bridges connected to each island must equal to the number written on that island.
2. *Noncrossing condition*: Each bridge cannot cross islands or other bridges.

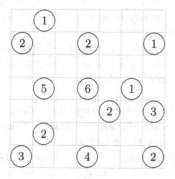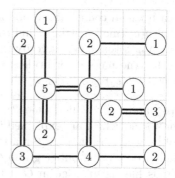

Fig. 3. An example of a Bridges puzzle (left) and its solution (right)

3. *Connecting condition*: The bridges must connect all islands into a single component.

Determining whether a given Bridges puzzle has a solution has been proved to be NP-complete [1]. Suppose P knows a solution of the puzzle and wants to convince V that it has a solution without revealing any information about the solution.

Define a *lip* to be a line segment of a unit length on the Bridges grid that either separates two adjacent cells or lies on the outer boundary of the grid. For each lip ℓ, let $b(\ell)$ be the number of bridges crossing through ℓ (including bridges coming out of the island from ℓ if ℓ is a lip of an island cell). First, P secretly places on ℓ a sequence encoding $b(\ell)$ in $\mathbb{Z}/3\mathbb{Z}$. Then, P publicly appends six ♣s to the end of the sequence to make it encode $b(\ell)$ in $\mathbb{Z}/9\mathbb{Z}$ (while ensuring V that $b(\ell)$ is at most 2). For each island cell c with a number $n(c)$, P publicly places a sequence encoding $n(c)$ in $\mathbb{Z}/9\mathbb{Z}$ on c.

For each cell c, let $b(\ell_1), b(\ell_2), b(\ell_3), b(\ell_4)$ be the numbers encoded by sequences on the top lip ℓ_1, the right lip ℓ_2, the bottom lip ℓ_3, and the left lip ℓ_4 of c, respectively (see Fig. 4). The steps of verifying P's solution of the puzzle are as follows.

$$
\begin{array}{c}
\ell_1 \\
\ell_4 \boxed{\; c \;} \ell_2 \\
\ell_3
\end{array}
$$

Fig. 4. Positions of lips $\ell_1, \ell_2, \ell_3, \ell_4$ surrounding a cell c.

1. For each lip ℓ located on the outer boundary of the Bridges grid, verify that $b(\ell) = 0$ (no bridge goes beyond the grid), which can be shown by simply revealing the sequence on ℓ.

2. For each island cell c with a number $n(c)$, verify that $b(\ell_1) + b(\ell_2) + b(\ell_3) + b(\ell_4) \equiv n(c) \pmod 9$ (the island condition).
3. For each non-island cell c, verify that $b(\ell_1) \equiv b(\ell_3) \pmod 9$ and $b(\ell_2) \equiv b(\ell_4) \pmod 9$ (the number of bridges passing through c is consistent), and also that $b(\ell_1) \cdot b(\ell_2) \equiv 0 \pmod 9$ (the noncrossing condition).

Steps 2 and 3 can be performed by applying a combination of copy and arithmetic protocols, which are explained in Appendix A, and the neighbor counting protocol in Sect. 2.6 (on a 2×9 matrix to verify the congruence).

Finally, construct a public graph G with all islands being vertices of G, and two islands having an edge in G if they are on the same row or column and there is no island between them (i.e. one can construct a valid bridge between them). Let H be a private subgraph of G such that two islands have an edge in H if there is at least one bridge between them in P's solution. P performs the following steps to commit H by placing a sequence $B(e)$, which is either $E_2(0)$ or $E_2(1)$, on every edge $e \in G$ to indicate whether $e \in H$.

1. For each edge $e \in G$ with endpoints u and v, consider any lip ℓ in the Bridges puzzle that lies between the two islands corresponding to u and v.
2. P picks the leftmost card on ℓ and places it as a leftmost card of $B(e)$ without revealing it.
3. P shuffles the second and third leftmost cards on ℓ and looks at the front side of them (without V seeing the front side). Then, P selects a ♣ among them and turns it over to reveal the front side to V. (If both cards are ♣s, P can select any of them; if only one card is a ♣, P must select it.)
4. P places another unselected card in Step 3 as a rightmost card of $B(e)$ without revealing it.

Observe that if there are one or two bridges between u and v, then $B(e)$ will be $E_2(1)$; if there is no bridge between them, then $B(e)$ will be $E_2(0)$. Hence, these steps ensure that the subgraph H is compatible with P's solution of the puzzle without revealing any information about it.

Verifying the connecting condition is equivalent to verifying that H is a spanning subgraph of G, which can be done by the protocol in Sect. 4. In total, this protocol uses $\Theta(pq)$ cards.

7   Future Work

We developed a physical card-based ZKP to verify the connected spanning subgraph condition, and showed applications of this protocol to verify solutions of three well-known NP-complete problems: the Hamiltonian cycle problem, the maximum leaf spanning tree problem, and the Bridges puzzle.

A possible future work is to explore methods to physically verify other NP-complete graph theoretic problems as well as other popular logic puzzles.

A   Copy and Arithmetic Protocols

In this appendix, we explain the copy and arithmetic protocols that can be used to verify problems in Sect. 6.

A.1   Copy Protocol

Given a sequence A encoding an integer a in $\mathbb{Z}/k\mathbb{Z}$, this protocol creates m additional copies of A without revealing a. It was developed by Shinagawa et al. [22].

1. Reverse the $k-1$ rightmost cards of A, i.e. move each $(i+1)$-th leftmost card of A to become the i-th rightmost card for $i = 1, 2, ..., k-1$. This modified sequence, called A', now encodes $-a \pmod{k}$.
2. Construct a $(m+2) \times k$ matrix M by placing the sequence A' in Row 1 and a sequence $E_k(0)$ in each of Rows $2, 3, ..., m+2$.
3. Apply the pile-shifting shuffle to M. Note that Row 1 of M now encodes $-a + r \pmod{k}$, and other rows now encode $r \pmod{k}$ for a uniformly random $r \in \mathbb{Z}/k\mathbb{Z}$.
4. Turn over all cards in Row 1 of M. Locate the position of a $\boxed{\heartsuit}$. Suppose it is at Column j.
5. Shift the columns of M cyclically to the left by $j-1$ columns. Turn over all face-up cards.
6. The sequences in Rows $2, 3, ..., m+2$ of M now encode $r - (-a+r) \equiv a \pmod{k}$, so we now have $m+1$ copies of A as desired.

A.2   Addition Protocol

Given sequences A and B encoding integers a and b in $\mathbb{Z}/k\mathbb{Z}$, respectively. This protocol computes the sum $a + b \pmod{k}$ without revealing a or b. It was developed by Shinagawa et al. [22].

1. Reverse the $k-1$ rightmost cards of A. This modified sequence, called A', now encodes $-a \pmod{k}$.
2. Construct a $2 \times k$ matrix M by placing A' in Row 1 and B in Row 2.
3. Apply the pile-shifting shuffle to M. Note that Row 1 and Row 2 of M now encode $-a+r \pmod{k}$ and $b+r \pmod{k}$, respectively, for a uniformly random $r \in \mathbb{Z}/k\mathbb{Z}$.
4. Turn over all cards in Row 1 of M. Locate the position of a $\boxed{\heartsuit}$. Suppose it is at Column j.
5. Shift the columns of M cyclically to the left by $j-1$ columns. Turn over all face-up cards.
6. The sequence in Row 2 of M now encodes $(b+r) - (-a+r) \equiv a+b \pmod{k}$ as desired.

A.3   Multiplication Protocol

Given sequences A and B encoding integers a and b in $\mathbb{Z}/k\mathbb{Z}$, respectively, this protocol computes the product $a \cdot b \pmod{k}$ without revealing a or b. It is a generalization of a protocol of Shinagawa and Mizuki [21] to multiply two integers in $\mathbb{Z}/3\mathbb{Z}$.

1. Repeatedly apply the copy protocol and the addition protocol to produce sequences $A_0, A_1, A_2, ..., A_{k-1}$ encoding $0, a, 2a, ..., (k-1)a \pmod{k}$, respectively.
2. Apply the sequence selection protocol to select the sequence A_b encoding $a \cdot b \pmod{k}$.

References

1. Andersson, D.: Hashiwokakero is NP-complete. Inf. Process. Lett. **109**(9), 1145–1146 (2009)
2. Bultel, X., Dreier, J., Dumas, J.-G., Lafourcade, P.: Physical zero-knowledge proofs for Akari, Takuzu, Kakuro and KenKen. In: Proceedings of the 8th International Conference on Fun with Algorithms (FUN), pp. 8:1–8:20 (2016)
3. Bultel, X., et al.: Physical zero-knowledge proof for Makaro. In: Izumi, T., Kuznetsov, P. (eds.) SSS 2018. LNCS, vol. 11201, pp. 111–125. Springer, Cham (2018). https://doi.org/10.1007/978-3-030-03232-6_8
4. Chien, Y.-F., Hon, W.-K.: Cryptographic and physical zero-knowledge proof: from sudoku to nonogram. In: Boldi, P., Gargano, L. (eds.) FUN 2010. LNCS, vol. 6099, pp. 102–112. Springer, Heidelberg (2010). https://doi.org/10.1007/978-3-642-13122-6_12
5. Dumas, J.-G., Lafourcade, P., Miyahara, D., Mizuki, T., Sasaki, T., Sone, H.: Interactive physical zero-knowledge proof for Norinori. In: Du, D.-Z., Duan, Z., Tian, C. (eds.) COCOON 2019. LNCS, vol. 11653, pp. 166–177. Springer, Cham (2019). https://doi.org/10.1007/978-3-030-26176-4_14
6. Garey, M.R., Johnson, D.S.: Computers and Intractability: A Guide to the Theory of NP-Completeness. W. H. Freeman & Co, San Francisco (1979)
7. Goldreich, O., Micali, S., Wigderson, A.: Proofs that yield nothing but their validity and a methodology of cryptographic protocol design. J. ACM **38**(3), 691–729 (1991)
8. Goldwasser, S., Micali, S., Rackoff, C.: The knowledge complexity of interactive proof systems. SIAM J. Comput. **18**(1), 186–208 (1989)
9. Gradwohl, R., Naor, M., Pinkas, B., Rothblum, G.N.: Cryptographic and physical zero-knowledge proof systems for solutions of sudoku puzzles. Theory Comput. Syst. **44**(2), 245–268 (2009)
10. Hashimoto, Y., Shinagawa, K., Nuida, K., Inamura, M., Hanaoka, G.: Secure grouping protocol using a deck of cards. IEICE Trans. Fund. Electron. Commun. Comput. Sci. **101.A**(9), 1512–1524 (2018)
11. Ibaraki, T., Manabe, Y.: A more efficient card-based protocol for generating a random permutation without fixed points. In: Proceedings of the 3rd International Conference on Mathematics and Computers in Sciences and Industry (MCSI), pp. 252–257 (2016)

12. Lafourcade, P., Miyahara, D., Mizuki, T., Sasaki, T., Sone, H.: A physical ZKP for Slitherlink: how to perform physical topology-preserving computation. In: Heng, S.-H., Lopez, J. (eds.) ISPEC 2019. LNCS, vol. 11879, pp. 135–151. Springer, Cham (2019). https://doi.org/10.1007/978-3-030-34339-2_8
13. Miyahara, D., et al.: Card-based ZKP protocols for Takuzu and Juosan. In: Proceedings of the 10th International Conference on Fun with Algorithms (FUN), pp. 20:1–20:21 (2020)
14. Miyahara, D., Sasaki, T., Mizuki, T., Sone, H.: Card-based physical zero-knowledge proof for Kakuro. IEICE Trans. Fund. Electron. Commun. Comput. Sci. **E102.A**(9), 1072–1078 (2019)
15. Nikoli: Hashiwokakero. https://www.nikoli.co.jp/en/puzzles/hashiwokakero.html
16. Robert, L., Miyahara, D., Lafourcade, P., Mizuki, T.: Interactive physical ZKP for connectivity: applications to Nurikabe and Hitori. In: Proceedings of the 17th Conference on Computability in Europe (CiE), pp. 373–384 (2021)
17. Robert, L., Miyahara, D., Lafourcade, P., Mizuki, T.: Physical zero-knowledge proof for Suguru puzzle. In: Devismes, S., Mittal, N. (eds.) SSS 2020. LNCS, vol. 12514, pp. 235–247. Springer, Cham (2020). https://doi.org/10.1007/978-3-030-64348-5_19
18. Ruangwises, S., Itoh, T.: Physical zero-knowledge proof for numberlink puzzle and k vertex-disjoint paths problem. N. Gener. Comput. **39**(1), 3–17 (2021)
19. Ruangwises, S., Itoh, T.: Physical zero-knowledge proof for ripple effect. In: Uehara, R., Hong, S.-H., Nandy, S.C. (eds.) WALCOM 2021. LNCS, vol. 12635, pp. 296–307. Springer, Cham (2021). https://doi.org/10.1007/978-3-030-68211-8_24
20. Sasaki, T., Miyahara, D., Mizuki, T., Sone, H.: Efficient card-based zero-knowledge proof for Sudoku. Theoret. Comput. Sci. **839**, 135–142 (2020)
21. Shinagawa, K., Mizuki, T.: Card-based protocols using triangle cards. In: Proceedings of the 9th International Conference on Fun with Algorithms (FUN), pp. 31:1–31:13 (2018)
22. Shinagawa, K., et al.: Card-based protocols using regular polygon cards. IEICE Trans. Fund. Electron. Commun. Comput. Sci. **E100.A**(9), 1900–1909 (2017)
23. Ueda, I., Miyahara, D., Nishimura, A., Hayashi, Y., Mizuki, T., Sone, H.: Secure implementations of a random bisection cut. Int. J. Inf. Secur. **19**(4), 445–452 (2019). https://doi.org/10.1007/s10207-019-00463-w

Non-instantaneous Information Transfer in Physical Reservoir Computing

Susan Stepney$^{(\boxtimes)}$ (ID)

Department of Computer Science, University of York, York, UK
susan.stepney@york.ac.uk

Abstract. The Echo State Network reservoir computing model is used in many applications. The original equations have a form of 'instantaneous' information flow. Here we suggest using a more physically realistic form of the equations, which takes into account the time needed for information to flow and to be processed. We demonstrate the effect of this change on a timeseries task, sunspot prediction, on a dynamical system emulation task, NARMA10, and on formulating 'reservoir of reservoir' equations.

1 Introduction

The Echo State Network (ESN) reservoir computer model, introduced in [9], has found wide application. In particular, it is used as a computational model for *in materio* reservoir computing. In its original form, it is a set of equations iterated in time in simulation. In its *in materio* form, it is used to model computation in real time physical devices. The original equations have a form of 'instantaneous' information flow, which becomes problematic when applied to physical devices.

Here we discuss the original form, suggest the use of a more physically realistic form, and evaluate the performances of each in two benchmark tasks.

2 The 'Instantaneous' Equations

2.1 The Original Form of the Reservoir Equations

The formulation of the reservoir equation typically given in the literature (see for example the original ESN formulation of [9, Eqs. 1, 2] and subsequent authors such as [4, Eqs. 22.1, 22.2], [18, Eqs. 1, 2]) is (translated into the notation used here):

$$\mathbf{x}(t) = f(\mathbf{W}\mathbf{x}(t-1) + \mathbf{W}_u\,u(t)) \tag{1}$$

$$v(t) = \mathbf{W}_v\,\mathbf{x}(t) \tag{2}$$

© Springer Nature Switzerland AG 2021
I. Kostitsyna and P. Orponen (Eds.): UCNC 2021, LNCS 12984, pp. 164–176, 2021.
https://doi.org/10.1007/978-3-030-87993-8_11

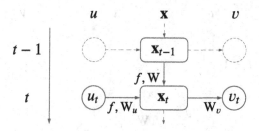

Fig. 1. A '(state)space-time' diagram, unfolding the recurrent structure, showing the information flow through the reservoir components. The input scalar, state vector, and output scalar state values are drawn on the horizontal axis; the discrete time steps are drawn on the vertical axis, with time progressing downwards. States are labelled with their discrete time values. The flow of information defined by Eqs. 1, 2 is shown by arrows, labelled with the required processing of that information through weight matrices and non-linear functions.

where $\mathbf{x}(t)$ is the state of the reservoir at time t, $\mathbf{W}$ is the reservoir weight matrix, $u(t)$ is the input at time t, $\mathbf{W}_u$ is the input weight matrix, f is a non-linear function (typically the tanh function), $v(t)$ is the output at time t, and $\mathbf{W}_v$ is the (trained) output weight matrix[1].

The equations define the state of the reservoir at current time t as a function of its state at the previous time $t-1$ and the current state of the input (Eq. 1), and the output at current time t as a function of the current state of the reservoir (Eq. 2). This is the typical formulation seen in the recurrent neural network literature in general, with an 'unfolded' time evolution as shown in Fig. 1.

However, the reader used to time-discretised differential equations, to difference equations, or to iterated maps, will notice the unusual use of the same time value on both sides of each of these equations. In those domains, the time-evolution equation is more typically expressed as $x(t+1) = \phi(x(t), x(t-1), \ldots)$. That is, the value of x at the new time $t+1$ is some function of values only at *earlier* times, not at the same time as in Eqs. 1, 2. This difference in formulation is the point examined in this paper.

2.2   The Issue: Apparent Instantaneous Information Flow

Equation 1 gives the state of the reservoir at time t in terms of the input value also at time t, despite the fact that the input value first needs to be processed through the input weight matrix $\mathbf{W}_u$ and the non-linear function f. Similarly, Eq. 2 gives the output value of the reservoir at time t in terms of the state of

---

[1] For simplicity, we assume that the input and output values are scalars, that the input scaling and spectral radius factors are folded into the respective weight matrices $\mathbf{W}_u$ and $\mathbf{W}$, and that there are no output feedback or leakiness terms. These assumptions do not affect the argument here, but do reduce notational clutter.

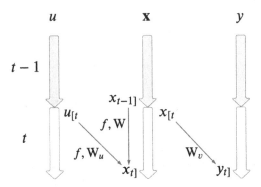

Fig. 2. A real-time interpretation of the original 'flat' reservoir equation that conforms to a non-instantaneous view. A time value '[t' indicates a value sampled the beginning of time unit t; a value 't]' indicates a value sampled at the end of time unit t.

the reservoir also at time t, despite the fact that the state value first needs to be processed through the output weight matrix W_v.

This can be illustrated using the (state)space-time diagram of Fig. 1. The issue is the existence of *horizontal* information flow arrows from input to state, and from state to output: this represents *instantaneous* information flow, which is unphysical, even before the need for processing that information through various weight matrices and non-linear functions is considered.

2.3   A Physically Reasonable Interpretation

It is possible to interpret the original formulation in a non-instantaneous manner, by taking one particular real-time systems viewpoint, where continuous time is divided into discrete time units as follows[2].

In this real-time systems view, because u is an input, it is sampled at the beginning of a time unit, whereas state $\mathbf{x}$ sampled at the end of a time unit. So Eq. 1 says that the state of the system at the *end* of time unit t, $\mathbf{x}(t)$, is some function of the state at the *end* of time unit $t-1$, $\mathbf{x}(t-1)$, plus the effect of the input at the *beginning* of time unit t, $u(t)$. So the actual time when u and $\mathbf{x}$ are sampled in order to calculate the update is in fact the same, because the end of time unit $t-1$ equals the beginning of time unit t. A similar argument can be applied to the output v equation.

This interpretation is summarised in Fig. 2; Eqs. 1, 2 are interpreted as:

$$\mathbf{x}(t]) = f(W\mathbf{x}(t-1]) + W_u\, u([t)) \tag{3}$$

$$\mathbf{v}(t]) = W_v\mathbf{x}([t) \tag{4}$$

---

[2] My thanks to David Griffin for suggesting this interpretation.

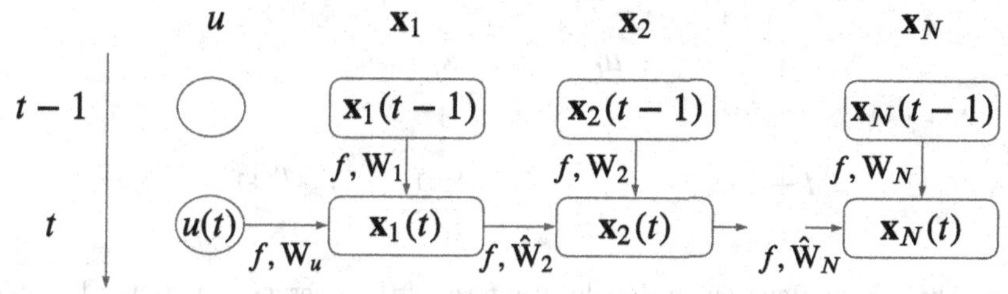

Fig. 3. The information flow in the 'deep ESN' architecture of [7].

Although this interpretation is physically plausible, removing the suggestion of instantaneous information flow and processing, it is hard to use mathematically when combining formulae and deriving results, because the symbol t in the original denotes different physical times, that of $[t$ and that of $t]$, in different contexts.

2.4 Pipelines of Reservoirs

As evidence that this potential interpretation is hard to use, or not even recognised, consider the case of the 'deepESN' layered reservoir architecture described in [7]. This architecture appears to be a pipeline of N reservoirs, the state of one acting as the input of the next.

Translating to the notation used here, and ignoring a leakiness term, the N layer model in [7, Eqs. 1, 2] is defined as:

$$\mathbf{x}_1(t) = f(W_1\mathbf{x}_1(t-1) + W_u\,u(t))$$
$$\mathbf{x}_i(t) = f(W_i\mathbf{x}_i(t-1) + \hat{W}_i\,\mathbf{x}_{i-1}(t)) \tag{5}$$

where $\mathbf{x}_i$ is the state in layer i; W_i is the weight matrix in layer i; $\hat{W}_i$ the weight matrix between layers $i-1$ and i. (We ignore the output for this discussion.) This has the same form as the single reservoir equation for each layer, with the state of the previous layer, $\mathbf{x}_{i-1}$, playing the role of the input to layer i.

The information flow defined by these equations is shown in Fig. 3. Here we see that the state of layer i at time t depends on the state of layer $i-1$ also at time t, and hence, propagating back, on the state of every prior layer and the input at time t. This is instantaneous transmission of the input and all prior layer states to layer N.

Despite superficial appearances, this is not a pipeline of reservoirs, but a more complicated system. This arrangement cannot be explained with the timing interpretation suggested above; instead, the instantaneous information transmission is a feature: "differently from the case of a standard ESN/RNN, the state information transmission between consecutive layers in a DeepESN presents no temporal delays" [6]. This setup is not just connecting the input to all layers simultaneously; it is also connecting the current state of one layer instantaneously to the current state of the next layer, despite first needing to feed it through weight matrix $\hat{W}$ and non-linear function f.

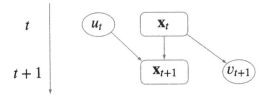

Fig. 4. The information flow in the physical form of the reservoir equations. The information flows only forwards (downwards) in time, with no instantaneous (horizontal) flows. (We omit the weight matrix labels from now on).

It might be possible to *simulate* this setup, by executing each layer individually in sequence within a single 'timestep', although the meaning of 'timestep' is then somewhat diluted. However, it is difficult to see how this could be physically implemented without providing a form of clocking internal to a timestep; this is, without providing finer grained timesteps. Additionally, this architecture cannot be adapted to recurrent inter-reservoir connections, that is, to a general 'reservoir of reservoirs' architecture, as a layer would find its state at time t is recurrently dependent on (a function of) its state at time t.

3   The 'Physical' Equations

3.1   Derived from Real-Time Interpretation

Consider the real-time interpretation of Fig. 2. We use the fact that the end of the one time interval coincides with the start of the next, $t] = [t + 1$, and substitute into Eqs. 3, 4, to get

$$\mathbf{x}(t + 1) = f(\mathbf{W}\mathbf{x}(t) + \mathbf{W}_u\, u(t)) \tag{6}$$

$$\mathbf{v}(t + 1) = \mathbf{W}_v\mathbf{x}(t) \tag{7}$$

where all the times refer to the same point of the time intervals. The information flow is summarised in Fig. 4. These equations are physically realisable, in that there is no instantaneous information flow, and also conform to the idea of difference equations, where the new state is defined in terms of past states only.

Several authors do in fact use a form similar to Eq. 6; see, for example, [5, Eq. 11], [12, Eq. 1], [14, Eqs. 1, 2], [16, p. 1165], [17, Eq. 2.1]. However, even these authors use a form like Eq. 2 rather than Eq. 7 for the output term.

3.2   Derived from Discretising Time in ODEs

Consider the two ordinary differential equations $\dot{x} = f(x, u)$, $\dot{v} = g(x)$, which describe a general open dynamical system where the state x evolves in time dependent on the current state, the input u, and the state function f; and where the output v evolves in time dependent on the current state and the output function g.

Discretise time, using the Euler method. The state evolution equation becomes:

$$\frac{x(t + \delta t) - x(t)}{\delta t} = f(x(t), u(t)) \tag{8}$$

Rearranging, putting $\delta t = 1$, and absorbing an $x(t)$ term by defining a new function $\hat{f}$, gives $x(t + 1) = \hat{f}(x(t), u(t))$. This has the same functional form as Eq. 6. In [11] there is a similar derivation of a leaky reservoir equation from an ODE; the appearance of $u(t)$ in the result, rather than the $u(t + 1)$ needed to match the form of Eq. 1, is described as due to "time indexing conventions".

Similarly discretising the output evolution gives $v(t + 1) = \hat{g}(x(t))$. This has the same functional form as Eq. 7. If the output evolution were not written as an ODE, but as $v(t) = g(x(t))$, it would not need to be discretised, and so would not result in two different times being involved. However, since some processing of the current state is required (given by g) in order to observe it, it arguably makes better physical sense to formulate the output as $v(t + 1) = g(x(t))$. Writing the output evolution explicitly as an ODE allows the function g to be more readily perceived as performing some action, requiring time.

3.3   Delay Line Reservoir Formulation

Appeltant *et al.* [1] investigate the use of the chaotic Mackey-Glass ODE [13, Eq. 4b] as the non-linear component in a delay line reservoir computer. For this purpose, they add an external input to the delayed feedback value, giving (in the notation used here):

$$\dot{x} = \frac{\beta(x(t - \tau) + \alpha u(t))}{1 + (x(t - \tau) + \alpha u(t))^n} - x(t) \tag{9}$$

where τ is the time delay, and α, β, n are adjustable parameters. Time discretising this equation gives the functional form $x(t + 1) = \phi(x(t), x(t - \tau), u(t))$.

4   Consequences of Using the 'Physical' Equations

Using the physical reservoir equations (Eqs. 6, 7), rather than the original equations (Eqs. 1, 2), has two main consequences discussed here. Firstly, it changes the measured performance of the reservoir (here discussed for time series tasks, and for dynamical system emulation tasks). Secondly, it eases the modelling of more complicated systems (here discussed for reservoirs of reservoirs).

	N	ρ	sparsity	runs	Twash	Ttrain	Ttest
sunspots	50	2	20%	50	320	500	2000
NARMA10	50	2	20%	50	100	500	2000

Fig. 5. Parameters for numerical experiments. N is the number of nodes in the reservoir; sparsity is the density of non-zero weights in $\mathbf{W}$ (uniformly distributed on $[-1, 1]$); ρ is the spectral radius; runs is the number of reservoirs tested per configuration; Twash is the number of timesteps in the initial washout period; Ttrain is the training period; Ttest is the testing period.

4.1 NRMSE Evaluation Measure

We use the normalised root mean square error (NRMSE) to evaluate different reservoir configurations. It is defined as [16, p. 1164]:

$$\mathrm{NRMSE} = \sqrt{\frac{\langle (\hat{v}_t - v_t)^2 \rangle}{\langle (\hat{v}_t - \langle \hat{v}_t \rangle)^2 \rangle}} \tag{10}$$

where $\hat{v}_t$ is the desired target output at time t, v_t is the actual output, and the averages $\langle \cdot \rangle$ are time averages.

The NRMSE does not depend on the scaling of the target $\hat{v}_t$. A perfect fit has NRMSE $= 0$. A value of NRMSE $= 1$ can be achieved by setting the reservoir output to a constant value equal to the time average of the target output, $v_t = \langle \hat{v}_t \rangle$ [11].

4.2 Time Series Prediction Task: Sunspots

In time series prediction, a reservoir is trained to predict the next (unseen) input in a given time series. One typical benchmark for time series prediction is the monthly sunspot count from 1749 to 1983, accessed from [3].

This sunspot dataset has 2820 entries. Here the data points are allocated sequentially to the washout, training, and testing periods as given in Fig. 5, and the values are normalised to lie in the range $[0, 0.5]$.

The performance of reservoirs on the sunspot prediction benchmark is examined for two different configurations (Fig. 6). The key requirement for the prediction task is that the target value $\hat{v}_t$ is equal to the next input value (as indicated by the information flowing 'backwards in time' in the target system. The experimental parameters are shown in Fig. 5.

The 'flat' reservoir configuration (Fig. 6a) has to predict one timepoint into the future: at time t the reservoir both sees input $u(t)$ and provides output $v(t)$ that is the predictor of $u(t+1)$. However, the 'physical' configuration (Fig. 6b) has to predict *three* timepoints into the future, due to the introduction of the input and output delays: at time t the reservoir sees input $u(t-1)$ and provides output $v(t+1)$ that is the predictor of $u(t+2)$. This suggests the physical reservoir will not be able to perform as well as the flat reservoir, as it has a more difficult task to perform.

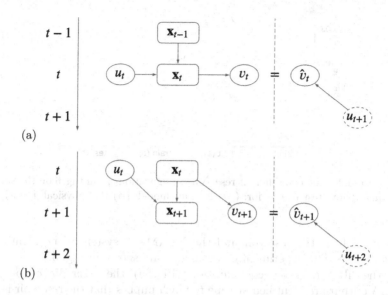

Fig. 6. Space-time diagrams of the reservoir equation (left) applied to the sunspot prediction benchmark (right), for different configurations of each. **(a)** The 'flat' form of the reservoir equation, where the input feeds instantaneously into the state and the state into the output. **(b)** The 'physical' form of the reservoir equations, where there are no instantaneous information flows.

The experimental results are shown in Fig. 7. As expected, the physical reservoir performs less well, although not disastrously so. (Alternatively, one might say the 'flat' configuration performs better than physically possible).

4.3 Dynamical System Emulation Task: NARMA10

Given an open (driven) dynamical system, the task is to train a reservoir to replicate its output (and hence emulate its dynamics) when driven with the same input.

The NARMA benchmark family was introduced with NARMA10 [2, Eq. 86]:

$$y(t+1) = 0.3y(t) + 0.05y(t)\sum_{i=0}^{9} y(t-i) + 1.5u(t-9)u(t) + 0.1 \quad (11)$$

The system has a single input $u(t)$; the task is to emulate the dynamics of this system and produce an output $y(t+1)$. For use as a benchmark, $u(t)$ is independent uniform noise in $[0, 0.5]$ [2, p. 705].

The performance of reservoirs against the NARMA10 benchmark[3] is examined for three different configurations (Fig. 8). The key requirement is that the

---

[3] The related NARMA20 benchmark [15] is defined with a tanh saturation function, in order to counteract its divergence. However, even NARMA10 used with the standard parameters can diverge [10]. In the experiments here, each generated target sequence is examined for divergence; a divergent sequence is discarded and a new sequence is calculated from a new input stream.

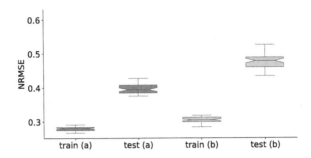

Fig. 7. Results for the two different reservoir configurations from Fig. 6 on the sunspot task. Training and test results for: **(a)** the 'flat' model; **(b)** the 'physical' model.

inputs u_t to both the reservoir and the NARMA10 system to be identical in value and time. The experimental parameters are shown in Fig. 5.

In the 'flat' reservoir configuration (Fig. 8a) the reservoir output and NARMA output are identified at time t, which implies that the reservoir is 'seeing' more input than is the NARMA system. Two 'physical' configurations are investigated. In the physical configuration of Fig. 8b, identifying outputs means that the reservoir is 'seeing' less input than is the NARMA system. So it would be expected to have a worse performance than the previous case. In the physical configuration of Fig. 8c, a corresponding output step, $\hat{v}(t+1) = y(t)$, is added to the NARMA system, allowing the reservoir and the NARMA system both to have the same information flow structure.

The experimental results are shown in Fig. 9. As expected, the physical reservoir that can see less input (Fig. 8b, Fig. 9b) performs less well than the 'flat' reservoir (Fig. 8a, Fig. 9a). However, the case where the physical reservoir emulates the target system adapted to have the same structure (Fig. 8c, Fig. 9c) performs best of all. This indicates that the structure of the task, as well as the structure of the reservoir, influences performance.

5   Further Formulations of the 'Physical' Equations

5.1   Incorporating Output Feedback

The full form of the equations from [9, Eqs. 1, 2] additionally include various feedbacks and other information flows between inputs and outputs (translated into the notation used here):

$$\mathbf{x}(t+1) = f(\mathbf{W}\mathbf{x}(t) + \mathbf{W}_u\, u(t+1) + \mathbf{W}_{back}v(t)) \tag{12}$$

$$v(t+1) = f_v(\mathbf{W}_v(\mathbf{x}(t+1), u(t+1), v(t))) \tag{13}$$

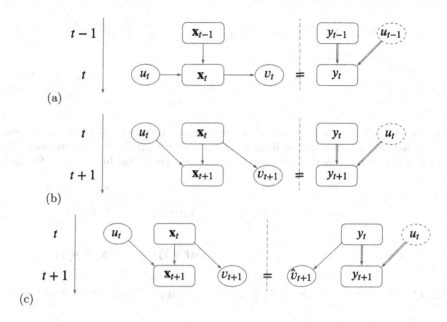

Fig. 8. Space-time diagrams of the reservoir equation (left) applied to the NARMA10 benchmark (right), for different configurations of each. **(a)** The 'flat' form of the reservoir equations. **(b)** The 'physical' form of the reservoir equations. **(c)** The physical form of the reservoir equations where the reservoir and target system have the same structure, achieved by adding an output $\hat{v}_t$ term to the latter.

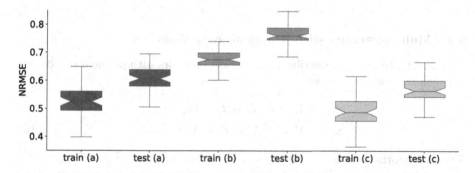

Fig. 9. Results for the three different reservoir configurations from Fig. 8 on the NARMA10 task. Training and test results for: **(a)** the 'flat' model; **(b)** the 'physical' model; **(c)** the physical model also adding an output to the NARMA10 system.

where W_{back} is a feedback weight matrix, and f_v is an output function. This has information flow as shown in Fig. 10a. Altering this to a physical system, where all information flows are forward in time, as shown in Fig. 10b, gives the physical form of the full equations:

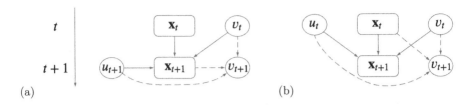

Fig. 10. The information flow in the full reservoir equations, with state equation flows as solid arrows, output equation flows as dashed arrows: (a) the 'flat' equations of [9, Eqs. 1, 2]; (b) the physical version with the same dependencies, but all flows forward in time, as Eqs. 14, 15.

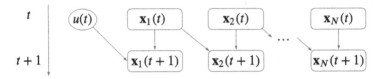

Fig. 11. The information flow in the physical reservoir pipeline architecture of Eq. 16.

$$\mathbf{x}(t+1) = f(W\mathbf{x}(t) + W_u\, u(t) + W_{back}v(t)) \tag{14}$$

$$v(t+1) = f_v(W_v(\mathbf{x}(t), u(t), v(t))) \tag{15}$$

5.2   Multi-reservoirs with 'Physical' Equations

We can rewrite Eq. 5 to describe a pipeline of reservoirs with physical timedelays between the components as:

$$\mathbf{x}_1(t+1) = f(W_1\mathbf{x}_1(t) + W_u\, u(t))$$
$$\mathbf{x}_i(t+1) = f(W_i\mathbf{x}_i(t) + \hat{W}_i\, \mathbf{x}_{i-1}(t)) \tag{16}$$

This has information flow as shown in Fig. 11 (again, ignoring the output).

We can use this physical formulation with more general connectivities, where sub-reservoir i may have (potentially recurrent) inputs from sub-reservoirs $j, k, \ldots$. If we assume that the weight matrices between unconnected reservoirs are zero, we can write:

$$\mathbf{x}_1(t+1) = f\left(W_u\mathbf{u}(t) + \sum_{1<j}\hat{W}_{1j}\mathbf{x}_j(t) + W_1\mathbf{x}_1(t)\right)$$

$$\mathbf{x}_i(t+1) = f\left(\sum_{j\neq i}\hat{W}_{ij}\mathbf{x}_j(t) + W_i\mathbf{x}_i(t)\right) \tag{17}$$

where $\hat{\mathbf{W}}_{ij}$ is the weight matrix connecting sub-reservoirs i and j. If sub-reservoirs are not directly, then that $\hat{\mathbf{W}}_{ij} = \mathbf{0}$.

This setup is equivalent to the single physical reservoir equation, with a weight matrix in block form: the diagonal blocks are the $\mathbf{W}_i$, and off-diagonal blocks are the (potentially less densely connected) $\mathbf{W}_{ij}$. Each component look like a reservoir (with a complicated input term), and the entire system looks like a reservoir. With the additional structure, we can adjust the various reservoir parameters (such as leak rate and spectral radius) on a per sub-reservoir basis, to tailor the system to be better suited to a range of more complex tasks. See also [14, Eqs. 10, 11].

With is physical approach, the sub-reservoirs can be connected in a variety of ways. Trained outputs can be added between components; connections between components can incorporate delays (for example, [8]), and other more complex configurations can be designed. The physical time components in the equations allow delays and information propagation to be readily analysed.

6 Conclusions

The original reservoir equations (Eqs. 1, 2), as typically used in the literature, are suitable for simulated single reservoir systems. However, their formulation makes them hard to apply to physical devices where information flow and processing takes time, and hard to manipulate mathematically to define cases where reservoirs are coupled together.

Here we use a 'physical' form of the equations (Eqs. 6, 7), where all flows, both from input to reservoir, and from reservoir to output, are explicitly forward in time. This form is demonstrated to have a slightly degraded performance when evaluated on a timeseries prediction task, but a potentially improved performance when evaluated on a dynamical system emulation task, if the structure of reservoir and task are matched. This demonstrates the importance of precisely specifying exactly how the task is setup and performed: space-time diagrams provide a clear visual way of doing so.

Additionally, the form of the physical equations makes is straightforward to manipulate and combine definitions of sub-reservoirs, since the time variable t denotes the same physical time wherever it appears.

Acknowledgments. Thanks to Matt Dale, David Griffin, Viv Kendon, Simon O'Keefe, Angelika Sebald, Martin Trefzer and Eleni Vasilaki for helpful conversations and for comments on a previous version. This work was funded by the SpInspired project, EPSRC grant EP/R032823/1, and the MARCH project, EPSRC grants EP/-V006029/1 and EP/V006339/1.

References

1. Appeltant, L., et al.: Information processing using a single dynamical node as complex system. Nat. Commun. **2**, 468 (2011). https://doi.org/10.1038/ncomms1476

2. Atiya, A.F., Parlos, A.G.: New results on recurrent network training: unifying the algorithms and accelerating convergence. IEEE Trans. Neural Networks **11**(3), 697–709 (2000). https://doi.org/10.1109/72.846741
3. Brownlee, J.: 7 time series datasets for machine learning. https://machinelearningmastery.com/time-series-datasets-for-machine-learning/. Accessed 21 Apr 2021
4. Dale, M., Miller, J.F., Stepney, S.: Reservoir computing as a model for *In-Materio* computing. In: Adamatzky, A. (ed.) Advances in Unconventional Computing. ECC, vol. 22, pp. 533–571. Springer, Cham (2017). https://doi.org/10.1007/978-3-319-33924-5_22
5. Dambre, J., Verstraeten, D., Schrauwen, B., Massar, S.: Information processing capacity of dynamical systems. Sci. Rep. **2**, 514 (2012). https://doi.org/10.1038/srep00514
6. Gallicchio, C., Micheli, A.: Deep echo state network (DeepESN): a brief survey. arXiv:1712.04323 [cs.LG] (2017)
7. Gallicchio, C., Micheli, A., Pedrelli, L.: Hierarchical temporal representation in linear reservoir computing. arXiv:1705.05782 [cs.LG] (2017)
8. Hihi, S.E., Bengio, Y.: Hierarchical recurrent neural networks for long-term dependencies. In: Touretzky, D.S., Mozer, M.C., Hasselmo, M.E. (eds.) Advances in Neural Information Processing Systems, vol. 8, pp. 493–499. MIT Press (1996)
9. Jaeger, H.: The "echo state" approach to analysing and training recurrent neural networks - with an erratum note. German National Research Center for Information Technology GMD Technical Report, Bonn, Germany, vol. 148, no. 34, p. 13 (2001)
10. Kubota, T., Nakajima, K., Takahashi, H.: A unifying framework for information processing in stochastically driven dynamical systems. arXiv:1906.04608 [cs.LG] (2019)
11. Lukoševičius, M.: A practical guide to applying echo state networks. In: Montavon, G., Orr, G.B., Müller, K.-R. (eds.) Neural Networks: Tricks of the Trade. LNCS, vol. 7700, pp. 659–686. Springer, Heidelberg (2012). https://doi.org/10.1007/978-3-642-35289-8_36
12. Lymburn, T., Jüngling, T., Small, M.: Quantifying robustness and capacity of reservoir computers with consistency profiles. In: Farkaš, I., Masulli, P., Wermter, S. (eds.) ICANN 2020. LNCS, vol. 12397, pp. 447–458. Springer, Cham (2020). https://doi.org/10.1007/978-3-030-61616-8_36
13. Mackey, M.C., Glass, L.: Oscillation and chaos in physiological control systems. Science **197**(4300), 287–289 (1977). https://doi.org/10.1126/science.267326
14. Manneschi, L., Ellis, M.O.A., Gigante, G., Lin, A.C., Del Giudice, P., Vasilaki, E.: Exploiting multiple timescales in hierarchical echo state networks. Front. Appl. Math. Stat. **6**, 76 (2021). https://doi.org/10.3389/fams.2020.616658
15. Rodan, A., Tiňo, P.: Minimum complexity echo state network. IEEE Trans. Neural Networks **22**(1), 131–144 (2011). https://doi.org/10.1109/TNN.2010.2089641
16. Schrauwen, B., Wardermann, M., Verstraeten, D., Steil, J.J., Stroobandt, D.: Improving reservoirs using intrinsic plasticity. Neurocomputing **71**(7), 1159–1171 (2008). https://doi.org/10.1016/j.neucom.2007.12.020
17. Seoane, L.F.: Evolutionary aspects of reservoir computing. Phil. Trans. R. Soc. B **374**(1774), 20180377 (2019). https://doi.org/10.1098/rstb.2018.0377
18. Tanaka, G., et al.: Recent advances in physical reservoir computing: a review. Neural Networks **115**, 100–123 (2019). https://doi.org/10.1016/j.neunet.2019.03.005

Quantum Logical Depth and Shallowness of Streaming Data by One-Way Quantum Finite-State Transducers (Preliminary Report)

Tomoyuki Yamakami[✉]

Faculty of Engineering, University of Fukui, 3-9-1 Bunkyo, Fukui 910-8507, Japan
TomoyukiYamakami@gmail.com

Abstract. The logical depth of a piece of data has served as a complexity measure to describe the amount of "usefulness" and "non-randomness" of information stored in the data itself, which is originated from Kolmogorov complexity. This notion of logical depth has been further expanded to various models of computation in the past literature to accommodate the needs for handling different computational circumstances. We focus on streaming data and streaming algorithms. With the use of one-way (or real-time) quantum finite-state automata equipped with write-once output tapes (called transducers) for recovering the desired information from the incoming compressed data sets, we introduce the notions of quantum finite-state depth and shallowness to capture the usefulness of the streaming data sets. We first layout a general setting of decompression of streaming data and, using its fundamental properties, we then argue the existence of deep and shallow data sets.

Keywords: Quantum finite-state automata · Quantum Kolmogorov complexity · Logical depth · Data compression · Data decompression · Quantum finite-state deep · Quantum finite-state shallow

1 Background, Motivations, and Challenges

The cultivation of efficient compression/decompression procedures of large-scale data sets has become an important subject in data science. Over the years, the methodology of algorithmic data compression/decompression has been extensively studied in computational complexity theory and information theory. *Theory of algorithmic information*, in particular, has arisen to clarify the minimum amount of algorithmically compressable/decompressable information and it has been widely applied to various other fields, such as physics, economy, mathematics, etc. The reader refers to, e.g., a textbook [16] for more explanations. To utilize underlying structures and patterns of available data sets, we first need to unearth them from the data set using a reasonable amount of computational resources (e.g., execution time, memory space, etc.). One technical challenge is

© Springer Nature Switzerland AG 2021
I. Kostitsyna and P. Orponen (Eds.): UCNC 2021, LNCS 12984, pp. 177–193, 2021.
https://doi.org/10.1007/978-3-030-87993-8_12

to seek for good methods of compressing data for the benefit of simple, efficient decompression procedures.

As a more concrete example, let us suppose that we have designed a piece of a computer program to efficiently search for an exact solution of a certain complicated real-life mathematical problem. If we have a chance to contact an external expert programmer for additional supportive information to help us achieve our task, what kind of data should we request? Should it be a random sequence, which is highly unstructured, or the sequence composed of all 1s, which contains high redundancy? Those two sequences are, nonetheless, not very useful for us in solving the problem in practice. We rather want even a small piece of meaningful, actually usable information on the desired solution so that we can effectively utilize such "useful" information to enhance the overall performance of our program. From a quantitative viewpoint, the random sequence and the sequence of all 1s respectively have very high and very low Kolmogorov complexities [15]; however, they turn out not to be useful in real-life practical applications. It is thus desirable for us to know how useful such information is and how much computational resources are necessary to unearth the meaningful portion out of the information. For this purpose, we need to quantitatively distinguish between useful data and not-useful data.

We therefore value "deep" information that exhibits high usefulness by simply dismissing both high randomness and high redundancy as "shallow" information. Since deep information could embody multiple layers of meaningful structures and patterns, we can unearth more useful information in general if we use more computational resources. In short, the more computational resources (such as time, space, etc.) we use, the more patterns we can discover. In contrast, shallow data contain relatively simple structures and patterns from which we cannot gain any more useful information even if we spend more execution time and memory space. One important feature of Bennett's *logical depth* notion is a fundamental property (known as a *slow growth law*) that no procedure can efficiently transform shallow information to deep information. The depth notion is therefore a technical tool in distinguishing between underlying complicated structures and simple structures hidden inside the given data. This notion considerably differs from Shannon's notion. In Shannon's theory, random sequences are considered to have a large amount of information, whereas Bennett's logical depth treats such random sequences less informative.

For the processing of off-line static data sets, a few variations of Bennett's depth notion have been proposed in the past literature for different complexity-theoretic applications. Juedes, Lathrop, and Lutz [13], for example, discussed the notion of *computational depth* and Antunes, Fortnow, van Melkebeek, and Vinodchandran [1] studied various other computational depth notions. Lately, Moser [17] also discussed *polylog depth*.

In this exposition, however, we are particularly interested in the compression/decompression of large *streaming data* coming incessantly through communication networks piece by piece. In processing such streaming data, a practical streaming algorithm needs to make a quick decision of what information to output as the data continue coming in piece after piece. The processing of such data requires quite different algorithmic approaches from the ones toward off-line static data sets on memory-loaded machines. For space-limited compression/decompression of streaming data, *one-way* (or *real-time*) *finite-state automata* have been widely acknowledged as a simple, practical model for implementing streaming algorithms with little memory. Historically, the notion of finite-state compressors/decompressors was introduced as early as in 1948 by Shannon [20] and later studied by Huffman [10] as well as Ziv and Lempel [28]. As another variant of Bennett's logical depth [3], Doty and Moser [8] used one-way deterministic finite-state automata as a computational model of decompression and introduced the notion of *finite-state depth*. Notice that Kolmogorov complexity (and thus Bennett's depth notion) relies on the existence of a universal Turing machine that can simulates all other Turing machines. Although no "universal" finite automaton exist, we can easily circumvent this problem by collectively considering *all* bounded-size finite-state automata. Lately, Jordon and Moser [11,12] considered its further extension to one-way pushdown automata and "Lempel-Ziv" algorithms [28]. Those automata-based formalisms of depth concept in [8,11,12] are therefore quite different from the Turing-machine-based formalism of [1,13,17].

It is quite natural to extend the notions of depth and shallowness of [8,11,12] to other models of compression/decompression procedures. In this exposition, we in particular intend to use *one-way quantum finite-state automata* and introduce the notions of *quantum finite-state depth* and *shallowness*. The idea of quantum-mechanical computation, nonetheless, dates back to the 1980s. Benioff [2], Deutsch [6,7], Yao [27], and Bernstein and Vazirani [4] proposed the early notions of quantum Turing machine and quantum circuit. See also [22]. Based on such quantum-mechanical machines, various quantum Kolmogorov complexity measures have been proposed in [5,9,21]. This exposition follows basic formalism of [8,11,12] analogously to quantum Kolmogorov complexity. The detailed definitions will follow in Sect. 2. In such a quantum setting, the major goal of our study is to (1) explore basic features of deep/shallow strings and sequences and (2) prove the existence of deep/shallow infinite sequences.

This work is merely an initial step toward a more exciting development of quantum logical depth and shallowness. We hope that its full fledged research will follow in the near future. See also Sect. 5 for a brief discussion.

2   Preparation: Basic Notions and Notation

2.1   Numbers, Sets, Languages, and Quantum States

We use the notations $\mathbb{N}$, $\mathbb{Z}$, and $\mathbb{C}$ to denote the sets of all *natural numbers* (i.e., nonnegative integers), of all integers, and of all complex numbers. Given a nonnegative constant e, the succinct notation $\mathbb{C}^{\leq e}$ expresses the set $\{\gamma \in \mathbb{C} : |\gamma| \leq e\}$. Given a finite set S, $|S|$ indicates the *cardinality* of S (i.e., the total number of elements in S). For two functions f and g on $\mathbb{N}$, we say that f *majorizes* g (denoted $f \geq g$) if $f(n) \geq g(n)$ holds for all $n \in \mathbb{N}$. We informally say that a property $P(n)$ holds *for almost all* $n \in \mathbb{N}$ if there exists a number $n_0 \in \mathbb{N}$

such that $P(n)$ holds for all $n \in \mathbb{N}$ with $n \geq n_0$. The notation $f + g$ denotes the function h defined as $h(x) = f(x) + g(x)$ for any input x. In particular, for any constant function $g(x) = c$, we succinctly write $f + c$. All *polynomials* are assumed to take nonnegative rational numbers. All *logarithms* are taken to the base 2. We set $\log^{(1)} x = \log x$ and $\log^{(k+1)} x = \log(\log^{(k)} x)$ for any $k \in \mathbb{N}^+$ and for any positive real number x. We also define $ilog(x)$ to be $\lceil \log x \rceil$ for $x > 0$ together with $ilog(0) = 0$.

We assume the reader's familiarity with the basics of formal languages and automata theory. In particular, λ denotes the *empty string* of length 0. For two strings (or two sequences) x and y, x is a *prefix* of y (or equivalently, y is a *suffix* of x), denoted $x \sqsubseteq y$, if there is a string (or a sequence) z such that $xz = y$. Given any alphabet Σ and any number $n \in \mathbb{N}$, Σ^n (resp., $\Sigma^{\leq n}$) denotes the set of all strings of length exactly n (resp., length at most n). We often identify decision problems with their associated languages.

Mostly, we use $\{0, 1\}$ as our alphabet. While the notation $\{0, 1\}^*$ denotes the set of all (finite-length) binary strings, we write $\{0, 1\}^\omega$ for the set of all infinite binary sequences. Let $s[i]$ denote the $(i + 1)$th symbol of s. It then follows that $s = s[0]s[1] \cdots s[|s| - 1]$. Given a sequence S, we use the notation $S\lceil n$ to denote the *initial segment* of S, consisting only of the first n elements of S. We also write $s\lceil n$ to the length-n prefix of a string s.

The reader's familiarity is also assumed with the basic quantum information and computation in, e.g., [14,18]. The notation $\||\phi\rangle\|$ denotes the ℓ_2-norm of $|\phi\rangle$, i.e., $\||\phi\rangle\| = \sqrt{\langle\phi|\phi\rangle}$. For any index $k \in \mathbb{N}^+$, $\mathcal{H}_k$ denotes a k-dimensional Hilbert space[1] and $\mathcal{H}_{all}$ is set to be $\bigcup_{k \geq 1} \mathcal{H}_k$. A *quantum bit* (or a qubit) is a unit-norm vector of $\mathcal{H}_2$. A unit (computational) basis of $\mathcal{H}_2$ is expressed as $\{|0\rangle, |1\rangle\}$. Given a quantum state $|\phi\rangle$ in a finite-dimensional Hilbert space, the notation $\ell(|\phi\rangle)$ denotes the logarithm of the dimension of this space. For instance, any vector $|\phi\rangle$ in $\mathcal{H}_{2^k}$ has length $\ell(|\phi\rangle) = k$.

2.2   One-Way Quantum Finite-State Transducers with Garbage Tapes

As noted in Sect. 1, our purpose is to model "streaming algorithms" that receive incoming streaming data, instantly process them, and produce the desired output values. After the incoming data are read from the input tape, they should be removed[2] (or erased) from the input tape.

Finite-state transducers are one of the simplest algorithmic procedures to transform strings to (possibly) different strings. A tape is called *write-once* if its tape head never moves to the left and, whenever the tape head writes a non-blank symbol, the tape head must move to the next blank cell. A core concept of this exposition is quantum finite-state transducers equipped with garbage

---

[1] It is possible to expand $\mathcal{H}_{2^k}$ to $\mathcal{H}_{\leq 2^k} = span\{|x\rangle \mid x \in \{0, 1\}^{\leq k}\}$, composed of basic vectors of mixed lengths.

[2] Unless we allow the 1qft to remove the read input data to a garbage tape, in the definition of QFS-complexity given later, $|\phi\rangle$ can be reduced to a classical string.

tapes [24, 26] (implicitly [23]). A *garbage tape* is a write-once tape onto which a quantum finite-state transducer dumps unnecessary information. The use of such a garbage tape is essential for quantum finite-state automata and, in particular, it helps the automata postpone any intermediate measurement until the end of their computation.

We use the following model of quantum finite-state transducers, which are used to transform any quantum state into another one. A *(one-way) quantum finite-state transducer* (or a 1qft, for short) T is a septuple $(Q, \Sigma, \{\vdash, \dashv\}, \Theta, \Gamma, \delta, q_0)$, where Σ is an *input alphabet*, Θ is a *garbage alphabet*, and Γ is an *output alphabet*, and q_0 ($\in Q$) is the initial (inner) state. The function δ is a *quantum transition function* $\delta : Q \times \check{\Sigma} \times Q \times \Theta^{\leq e} \times \Gamma^{\leq e} \to \mathbb{C}^{\leq 1}$ for a constant $e \in \mathbb{N}^+$, where $\check{\Sigma} = \Sigma \cup \{\vdash, \dashv\}$. No halting state is needed because M's tape head must halt when reaching the right endmarker $\dashv$. Each output value of δ is called a *transition amplitude* or simply an *amplitude*. Generally, we allow arbitrary complex amplitudes for 1qft's. For readability, we intentionally write $\delta(p, \sigma | q, \theta, \tau)$ instead of $\delta(p, \sigma, q, \theta, \tau)$. The number e indicates the maximal length of string that can be written down on an output tape as well as a garbage tape in a single step and this value e is called the *production size* of T, denoted $prod(T)$.

The transducer T begins to scan the left endmarker $\vdash$ in the initial (inner) state $|q_0\rangle$ and then halts after processing the right endmarker $\dashv$. All tape cells are indexed by natural numbers. For the input tape, cell 0 contains $\vdash$ and cell $|x| + 1$ contains $\dashv$ if x is an input string. The transducer T moves its input-tape head in one direction from left to right whenever it scans a tape symbol. Such a movement is also referred to as "real time." Similarly, the two other heads of garbage and output tapes move in one direction to the next blank cells whenever they write non-blank symbols. A very important point is that, whenever M reads an input symbol σ, M must *erase* σ (namely, replace σ by a designated blank symbol B). This process can be carried out "unitarily" because of the existence of the garbage and the output tapes. Assuming that T is in inner state p scanning symbol σ, if we apply a transition of the form $\delta(p, \sigma | q, \theta, \tau) = \alpha$, then T changes its inner state to q, writes θ on its garbage tape, and writes τ on its output tape with amplitude α. In particular, when either $\theta = \lambda$ or $\tau = \lambda$, we interpret this step as a *stationary move* (or a *λ-move*) of T; namely, T's tape head stays still with writing no non-blank symbol. A *configuration* of M is a quadruple (q, y, z, w), which indicates that M is in inner state q, scanning the first symbol of the remaining input string y, the garbage tape consists of z, and the output tape (except for the blank cells) contains w. A single application of δ changes one configuration to another. When an input x is given, the initial configuration is of the form $(q_0, \tilde{x}, \lambda, \lambda)$ and a final configuration has one of the forms (q, λ, z, w), where $\tilde{x} = \vdash x \dashv$. A series of configurations from an initial configuration to a final configuration is a *computation path* if each intermediate configuration is obtained from the previous one by a single application of δ. It is important to remember that M is always assumed to behave as a *unitary operator* over the Hilbert space of its configurations (which is called a *configuration space*).

A *(projective) measurement* in a computation basis $\mathcal{B}$ is a projection Π of any given quantum state onto the Hilbert space spanned by all elements of

$\mathcal{B}$. After M halts with a quantum state $|\xi\rangle$, we perform a measurement and obtain a classical string y with probability $\|\langle\xi|y\rangle\|^2$, where $\langle\xi|y\rangle$ means the un-normalized quantum state obtained from $|\xi\rangle$ after observing y. See [25] for its formal treatment. We define the *projective majority function* τ as $\tau(|\xi\rangle) = y$ iff $\|\langle\xi|y\rangle\|^2 > \frac{1}{2}$, provided that such a string y exists. Notice that this string y is uniquely determined. This y is called the *projective majority string* in $|\xi\rangle$. For brevity, we say that, on input $|\phi\rangle \in \mathcal{H}_{\text{all}}$, T *outputs y with probability η* if T starting with $|\phi\rangle$ given on the input tape with the two endmarkers produces y on its output tape with probability η. Moreover, we denote by $\ell_{\text{opt}}(|\xi\rangle)$ the length of output strings y observed in $|\xi\rangle$.

A "natural" complexity measure for a 1qft T is the *state complexity* of T, denoted $state(T)$, which is the number $|Q|$ of all inner states. Another important measure is the aforementioned production size of T.

For comparison, we also consider a deterministic analogue of 1qft's. A *one-way deterministic finite-state transducer* (or a 1dft) is of the form $(Q, \Sigma, \{\vdash, \dashv\}, \Gamma, \delta, q_0)$ with a map δ from $Q \times \check{\Sigma}$ to $Q \times \Gamma^{\leq e}$.

2.3   Description Sizes of 1dft's and 1qft's

Recall from Sect. 1 that there is no "universal" 1qft that can simulate any other 1qft. To circumvent the lack of universality, we instead consider a family of all 1qft's of bounded "sizes" that can produce a target string x with high probability from certain compressed quantum information $|\phi\rangle$. Our intention is to define the decompression complexity in Sect. 3.1. For our purpose, we need to properly define the "size" of each 1qft.

We begin with an easy case of a 1dft T. Since T's behavior is completely dictated by δ, we first express δ as a "table", indexed by the elements in $Q \times \check{\Sigma}$, in which all entries are certain elements in $Q \times \Gamma^{\leq e}$, where e is the production size of T. The number of such tables is 2^t for $t = |Q||\check{\Sigma}|(\log|Q||\Gamma^{\leq e}|)$, which equals $4|Q|(\log|Q| + e + 1)$. Since the table size depends on the state complexity $state(T)$ and the production size e, we introduce the *description size* of T as $4|Q|(i\log|Q| + e + 1)$. The notation $|T|$ is used to denote the description size of T. For each constant $k \in \mathbb{N}^+$, the notation $\text{DFST}^{\leq k}$ expresses the set of all 1qft's T whose description size $|T|$ is at most k. It follows that $|\text{DFST}^{\leq k}| \leq 2^{k+1}$.

As for 1qfa's, an effective encoding of each 1qfa was given in [19,26] using a quantum-circuit approximation scheme for its time-evolution operator. However, we here do not want to rely on such a complicated scheme. In a similar way to the case of 1dft's, T's *time-evolution operator*[3] can be indexed by two sets $Q \times \check{\Sigma}$ and $Q \times \Theta^{\leq e} \times \Gamma^{\leq e}$. For each index $(q, \sigma) \in Q \times \check{\Sigma}$, let us consider a quantum state $|q, \sigma\rangle \otimes |\psi_{q,\sigma}\rangle$ for $|\psi_{q,\sigma}\rangle = \sum_{p,\theta,\tau} \delta(q, \sigma|p, \theta, \tau)|p, \theta, \tau\rangle$. We further set $|\Psi_T\rangle = \bigotimes_{q,\sigma} |\psi_{q,\sigma}\rangle$, where (q, σ) ranges over $Q \times \check{\Sigma}$. This quantum state $|\Psi_T\rangle$ clearly "describes" T and it has dimension 2^t for $t = |Q||\check{\Sigma}||Q||\Theta^{\leq e}||\Gamma^{\leq e}|$. Since

---

[3] This unitary operator, which is naturally induced from δ, transforms a superposition of configurations to another one.

t equals $4|Q|^2(e+1)^2$, we define the *description size* $|T|$ of T to be $4|Q|^2(e+1)^2$. Similar to $\mathrm{DFST}^{\leq k}$, $\mathrm{QFST}^{\leq k}$ is defined in terms of 1qft's instead of 1dft's.

3 Basic Definitions

3.1 Quantum Finite-State Compressors and Decompressors

The notion of quantum finite-state transducers is a natural extension of finite-state transducers. In this exposition, we thus use 1qft's to compress and decompress streaming data sets in a quantum-mechanical way. Note that, since a 1qft T behaves in a quantum fashion, it can produce a superposition of different outputs. For readability, we call a 1qft a *QFS-compressor* (resp., a *QFS-decompressor*) if it is used to compress (resp., decompress) given data sets.

Fix our input alphabet Σ to be $\{0,1\}$. Instead of considering the minimal size of programs on a single "universal" Turing machine, we consider all possible QFS-decompressors for defining an appropriate decompression complexity. Let us introduce a key notion, $\mathrm{QDC}_\varepsilon^p(x)$, which measures the minimal size of inputs from which 1qft's can produce target strings x with reasonably high probability. Let $\varepsilon \in [0,1)$ be a fixed error bound. The (p,ε)-*quantum-finite-state decompression complexity*[4] (or the (p,ε)-*QFSD-complexity*) of a string x, denoted by $\mathrm{QDC}_\varepsilon^p(x)$, is defined as $\min_T \{\mathrm{QDC}_\varepsilon(T:x) \mid T \in \mathrm{QFST}^{\leq p(|x|)}\}$, where $\mathrm{QDC}_\varepsilon(T:x) = \min_{|\phi\rangle,T} \{\ell(|\phi\rangle) \mid |\phi\rangle \in \mathcal{H}_{\mathrm{all}} \wedge T(|\phi\rangle) = |\xi\rangle \wedge \ell_{\mathrm{opt}}(|\xi\rangle) = |x| \wedge \|\langle\xi|x\rangle\|^2 \geq 1-\varepsilon\}$. Intuitively, a 1qft T of description size at most $p(|x|)$ takes a compressed quantum state $|\phi\rangle$ in $\mathcal{H}_{\mathrm{all}}$ as its input and transforms it to $|\xi\rangle$ whose outputs must be close to $|x\rangle$. As before, when $p(n) = k$ for a constant $k \in \mathbb{N}$, we write $\mathrm{QDC}_\varepsilon^k(x)$. Since $\varepsilon \neq 1$, for later use, when we only demand that $\|\langle\xi|x\rangle\|^2 > 0$, we explicitly write $\mathrm{QDC}_1(T:x)$ and $\mathrm{QDC}_1^p(x)$. Obviously, $\mathrm{QDC}_\varepsilon^k(x) \geq \mathrm{QDC}_1^k(x)$ holds for any $\varepsilon \in [0,1)$. Since every step of T in $\mathrm{QFST}^{\leq p(|x|)}$ produces at most $e = prod(T)$ symbols, it follows that $\mathrm{QDC}_\varepsilon^p(x) \geq \frac{|x|}{e}$.

As a quick example, let us consider a unique identity machine M_{id}, which is a 1qft M_{id} that, on input z, produces z on its output tape. If k_0 denotes the description size of M_{id}, then we obtain $\mathrm{QDC}_0^{k_0}(z) \leq |z|$ for all $z \in \Sigma^*$.

A simple *majority vote strategy* is often taken to amplify the probability of successfully producing the target string x. As shown below, such a strategy over, say, 3 samplings may increase the QFST-complexity 3 folds.

The *shuffled string* of three given strings z_1, z_2, z_3, denoted $[z_1, z_2, z_3]$, is defined as follows. Let $z_1 = a_1 a_2 \cdots a_k$, $z_2 = b_1 b_2 \cdots b_k$, and $z_3 = c_1 c_2 \cdots c_k$. The shuffled string $[z_1, z_2, z_3]$ is then set to be $a_1 b_1 c_1 a_2 b_2 c_2 \cdots a_k b_k c_k$.

Lemma 1. *Let $\varepsilon \in [0, 1/2)$ and $\varepsilon' = \varepsilon^2(3 - \varepsilon)$. There exists a constant $c > 0$ such that $\mathrm{QDC}_{\varepsilon'}^{p+c}(x) \leq 3\mathrm{QDC}_\varepsilon^p(x)$ for any x.*

---

[4] Following Vitányi's formulation [21], however, it is also possible to define $\mathrm{QDC}^p(x) = \min_{T,|\phi\rangle} \{\ell(|\phi\rangle) + \lceil -\log \|\langle\xi|x\rangle\|^2 \rceil : T \in \mathrm{QFST}^{\leq p(|x|)}, T(|\phi\rangle) = |\xi\rangle \wedge \ell_{\mathrm{opt}}(|\xi\rangle) = |x|\}$.

Proof. Let $k = \mathrm{QDC}^p_\varepsilon(x)$ and take $T \in \mathrm{QFST}^{\leq p(|x|)}$ and $|\phi\rangle, |\xi\rangle \in \mathcal{H}_{\mathrm{all}}$ such that $T(|\phi\rangle) = |\xi\rangle$, $\ell_{\mathrm{opt}}(|\xi\rangle) = |x|$, and $\|\langle \xi | x \rangle\|^2 \geq 1 - \varepsilon$. Let $|\phi\rangle = \sum_{z:|z|=k} \alpha_z |z\rangle$ with $\sum_{z:|z|=k} |\alpha_z|^2 = 1$. Consider $|\phi^{(3)}\rangle = \sum_{z_1, z_2, z_3} \alpha_{z_1} \alpha_{z_2} \alpha_{z_3} |[z_1, z_2, z_3]\rangle$.

We design a new 1qft M that reads $[z_1, z_2, z_3]$ in $|\phi^{(3)}\rangle$ and simulates $T(\rho_1)$, $T(\rho_2)$, and $T(\rho_3)$ in parallel, where ρ_i is obtained by tracing out all z_j for $j \neq i$. If at least two simulations reach accepting states, then M accepts; otherwise, M rejects. The total error probability of M on $|\phi^{(3)}\rangle$ is at most $\sum_{i=2}^3 \binom{3}{i} \varepsilon^{3-i}(1 - \|\langle \xi_i | x \rangle\|^2)^i \leq \varepsilon^2(3 - \varepsilon) = \varepsilon'$. $\square$

A *one-way reversible finite transducer* (or a 1rft, for short) can be seen as a special case of 1qft restricted to classical computation with no garbage tape and no error. A function $f : \{0,1\}^\omega \to \{0,1\}^\omega$ is called *RFS-computable* if there exist a 1rft T and a series $\{x_n\}_{n \in \mathbb{N}}$ such that, for all sequences $S \in \{0,1\}^\omega$, (i) $T(S\lceil n) = x_n$ for any $n \in \mathbb{N}$, (ii) $\lim_{n \to \infty} |x_n| = \infty$, and (iii) $x_n \sqsubseteq f(S)$ for any $n \in \mathbb{N}$. Note that, since T is one-way, $|x_n| \leq |x_{n+1}|$ immediately follows. For readability, we succinctly write $T(S) = f(S)$ if f is RFS-computable by T.

3.2   Quantum Finite-State Depth and Shallowness

Let us introduce the notion of *quantum finite-state depth/shallowness* in details. We are more concerned with the difference between two values $\mathrm{QDC}^{p_1}_\varepsilon(x)$ and $\mathrm{QDC}^{p_2}_\varepsilon(x)$ for two separate functions p_1 and p_2. For such functions p_1 and p_2, assuming that $p_1 \leq p_2$, we define the complexity measure $\mathrm{QD}^{p_1,p_2}_\varepsilon(x)$ to be the difference $\mathrm{QDC}^{p_1}_\varepsilon(x) - \mathrm{QDC}^{p_2}_\varepsilon(x)$ for any $x \in \Sigma^*$.

The depth and shallow notions are more suited for an application to infinite sequences than individual finite strings. We are thus more interested in infinite sequences and their QFS-decompression complexity. Given an infinite sequence $S \in \{0,1\}^\infty$, we say that S is $(\mu, \varepsilon, \mathcal{G})$-*quantum-finite-state deep* (or $(\mu, \varepsilon, \mathcal{G})$-QFS-deep) if, for any $p \in \mathcal{G}$, there exists an appropriate function $q \in \mathcal{G}$ with $q \geq p$ such that $QD^{p,q}_\varepsilon(S\lceil n) \geq \mu(n)$ holds for almost all numbers n in $\mathbb{N}$. Furthermore, S is said to be $(\mu, \varepsilon, \mathcal{G})$-*quantum-finite-state shallow* (or $(\mu, \varepsilon, \mathcal{G})$-QFS-shallow) if there exists a $p \in \mathcal{G}$ such that, for any function $q \in \mathcal{G}$ with $q \geq p$, $\mathrm{QD}^{p,q}_\varepsilon(S\lceil n) \leq \mu(n)$ holds for almost all n in $\mathbb{N}$. Notice that the above two notions, $(\mu, \varepsilon, \mathcal{G})$-QFS-depth and $(\mu, \varepsilon, \mathcal{G})$-QFS-shallowness, are *not* logically opposite to each other. Given two sets $\mathcal{M}$ and $\mathcal{G}$, we further say that S is $(\mathcal{M}, \mathcal{G})$-*QFS-deep* (resp., $(\mathcal{M}, \mathcal{G})$-QFS-shallow) if there are elements $\mu \in \mathcal{M}$ and $\varepsilon \in [0, 1/2)$ such that S is $(\mu, \varepsilon, \mathcal{G})$-QFS-deep (resp., $(\mu, \varepsilon, \mathcal{G})$-QFS-shallow).

As candidates of the aforementioned two sets $\mathcal{M}$ and $\mathcal{G}$, we will consider the following three sets of specific functions: $\mathrm{lin} = \{p \mid p$ is a linear function$\}$, $\log = \{f \mid f$ is a logarithmic function$\}$, and $\mathrm{const} = \{f \mid f$ is a constant function$\}$.

In comparison, we also define a deterministic analogue of one-way quantum finite-state depth and shallowness. With the use of 1dft's, an FS analogue of $\mathrm{QDC}^p_\varepsilon(x)$ is formulated as follows. Let $\mathrm{DDC}^p(x) = \min\{\mathrm{DDC}(T : x) \mid T \in \mathrm{DFST}^{\leq p(|x|)}\}$, where $\mathrm{DDC}(T : x) = \min\{|w| : T(w) = x\}$. Finally, we define the *one-way deterministic finite-state depth complexity* as $\mathrm{DD}^{p_1,p_2}(x) = \mathrm{DDC}^{p_1}(x) - \mathrm{DDC}^{p_2}(x)$. When p is a constant function of the form $p(n) = k$ for all $n \in \mathbb{N}$, we write $\mathrm{DDC}^k(x)$ in place of $\mathrm{DDC}^p(x)$.

Lemma 2. *Let k and n be any two numbers in $\mathbb{N}^+$. Assume that $k \leq n+1$ for all n. There exists a string $x \in \Sigma^n$ for which $\mathrm{DDC}^k(x) \geq n - k - 1$.*

Proof. Let $\mu(n) = \max\{\mathrm{DD}^c(x) \mid x \in \Sigma^n\}$. Consider a map from x in Γ^n to (w,T) in $\Sigma^{\mu(n)} \times \mathrm{DFST}^{\leq k}$ such that $T(w) = x$. By this map, we obtain $2^n \leq 2^{\mu(n)} \cdot 2^{k+1}$. This is equivalent to $n \leq \mu(n) + k + 1$. Therefore, it follows that $\mu(n) \geq n - k - 1$. □

The notion of $(\mathcal{M}, \mathcal{G})$-*DFS-depth* is defined in a similar way to the $(\mathcal{M}, \mathcal{G})$-QFS-depth. This notion will be used in Proposition 15.

4 Major Contributions

4.1 Basic Properties of QFSD-Complexity

We start with studying fundamental properties of QFS-complexity. Berthiaume et al. [5] claimed that the quantum Kolmogorov complexity of classical strings coincides with the classical one modulo a certain constant.

Since, in our 1qft setting, quantum computation can simulate deterministic computation with the help of garbage tapes, we obtain the following simple relationship between DFS- and QFS-complexities.

Lemma 3. *There exists a constant $c > 0$ such that, for any function p, $\mathrm{DDC}^p(x) \geq \mathrm{QDC}_0^{p+c}(x)$ holds for all strings $x \in \Sigma^*$.*

Proof. Let $k = \mathrm{DDC}^p(x)$ and take $w \in \Sigma^*$ and $M \in \mathrm{DFST}^{\leq p(|x|)}$ such that $M(w) = x$ and $k = |w|$. Since M is a 1dft, we can transform it to an "equivalent" 1qft, say, T that simulates M with probability 1 properly using a garbage tape. Note that $state(T) = state(M) + O(1)$. Therefore, T is in $\mathrm{QFST}^{\leq p(|x|)+c}$ for a certain constant $c > 0$. By the definition, it follows that $T(|w\rangle) = |\xi\rangle$, $\ell_{\mathrm{opt}}(|\xi\rangle) = |x|$, and $\|\langle\xi|x\rangle\|^2 = 1$. This implies that $\mathrm{QDC}_0^{p+c}(x) \leq k$. □

In what follows, we naturally expand a classical string x used in the definition of $\mathrm{QDC}_\varepsilon^p(x)$ to a quantum state $|\psi\rangle$.

Theorem 4. *Fix $k, m, n \in \mathbb{N}^+$ with $k + m + 2 < n$. There exist a subspace V of $\mathcal{H}_{2^n}$ of dimension $2^n - 2^{k+m+2}$ and a string $x \in \Sigma^n$ satisfying $\mathrm{QDC}_1^k(|\psi\rangle) \geq m + 1$ for any $|\psi\rangle \in V$.*

Proof Sketch. For convenience, we write $\mathcal{H}_{\leq 2^m}$ for $\bigoplus_{i=1}^m \mathcal{H}_{2^i}$. In this proof, we view a description of 1qft in $\mathrm{QFST}^{\leq k}$ as a vector in $\mathcal{H}_{\leq 2^k}$ and view $\mathrm{QFST}^{\leq k}$ as the Hilbert space spanned by all such vectors. Let $\mathcal{P}_{k,m}$ denote the Hilbert space spanned by appropriate basis vectors of $\mathrm{QFST}^{\leq k} \otimes \mathcal{H}_{\leq 2^m}$. The dimension of $\mathcal{P}_{k,m}$ is $2^{k+1} \cdot 2^{m+1} = 2^{k+m+2}$.

We consider a map U defined by $U(|T\rangle|\phi\rangle) = T(|\phi\rangle)$. We define A as the set of all vectors in $\mathcal{H}_{2^n}$ that are orthogonal to all vectors of the form $U(|T\rangle|\phi\rangle)$ for $|T\rangle|\phi\rangle \in \mathcal{P}_{k,m}$. Note that the dimension of $U(|T\rangle|\phi\rangle)$ is 2^n.

Finally, we choose the set V of all elements in $\mathcal{H}_{2^n}$ that are orthogonal to all vectors in A. This set V has dimension $2^n - 2^{k+m+2}$. It clearly follows that $\mathrm{QDC}_1^k(|\psi\rangle) \geq m + 1$ for all vectors $|\psi\rangle \in V$.   □

Corollary 5. *Let* $k \in \mathbb{N}^+$ *and* $\varepsilon \in [0, 1/2)$. *For any sufficiently large* n, *there exists a string* $x \in \Sigma^n$ *such that* $\mathrm{QDC}_\varepsilon^k(x) \geq n - k - 2$.

Proof. Let n be any sufficiently large integer. We set $m = n - k - 3$, which satisfies $k + m + 2 < n$. It follows by Theorem 4 that there exists a string $x \in \Sigma^n$ for which $\mathrm{QDC}_\varepsilon^k(x) \geq \mathrm{QDC}_1^k(x) \geq m + 1 = n - k - 2$.   □

Next, we show an *invariance property* of QFS-complexity under certain information lossless operations. A weak form of the so-called *slow growth law* holds for our notion of QFS-deep information.

Theorem 6. *Let* S *be any sequence in* $\{0,1\}^\omega$ *and let* $f : \{0,1\}^\omega \to \{0,1\}^\omega$ *be a RFS-computable. If* $f(S)$ *is* $(\mathcal{M}, \mathcal{G})$-*QFS-deep, then* S *is also* $(\mathcal{M}, \mathcal{G})$-*QFS-deep.*

To verify the theorem, we first prove the following general lemma. Let us recall that $\tau(|\xi\rangle)$ denotes the projective majority string in $|\xi\rangle$.

Lemma 7. *Let* $\varepsilon, \varepsilon' \in [0, 1/2)$. *Let* M *be a 1qft with error-probability at most* ε' *and let* N *be a 1rft. Let* p *be any function on* $\mathbb{N}$. *Assume that* $\tilde{\varepsilon} \geq \varepsilon + \varepsilon' - \varepsilon\varepsilon'$.

1. *There exists a* $c > 0$ *such that* $\mathrm{QDC}_\varepsilon^p(x) \geq \mathrm{QDC}_{\tilde{\varepsilon}}^{p+c}(\tau(M(x)))$ *for any* x.
2. *There exists a* $c > 0$ *such that* $\mathrm{QDC}_\varepsilon^p(N(x)) \geq \mathrm{QDC}_\varepsilon^{p+c}(x)$ *for any* x.

Proof Sketch. (1) Let $M = (\hat{Q}, \Sigma, \{\vdash, \dashv\}, \hat{\Theta}, \hat{\Gamma}, \hat{\delta}, \hat{q}_0)$ be given as in the lemma and let $y = \tau(M(x))$. Let $m = \mathrm{QDC}_\varepsilon^p(x)$. Take $|\phi\rangle, |\xi\rangle \in \mathcal{H}_{\mathrm{all}}$ and $T \in \mathrm{QFST}^{\leq p(|x|)}$ such that $\ell(|\phi\rangle) = m$, $T(|\phi\rangle) = |\xi\rangle$, $\ell_{\mathrm{opt}}(|\xi\rangle) = |x|$, and $\|\langle\xi|x\rangle\|^2 \geq 1 - \varepsilon$. Assume that T has the form $(Q_T, \Sigma, \{\vdash, \dashv\}, \Theta_T, \Gamma_T, \delta_T, q_{0,T})$, provided that $Q_T \cap \hat{Q} = \varnothing$, $\Theta_T \cap \hat{\Theta} = \varnothing$, and $\Gamma_T \cap \hat{\Gamma} = \varnothing$.

We wish to define a new 1qft T' to produce y. Intuitively, on input $|\phi\rangle$, we run T on $|\phi\rangle$ and, instead of writing symbols on the output tape, we feed them to M and run it. This is possible because T outputs at most a fixed number of symbols at each step and we can run M on these symbols incessantly. Let $|\xi\rangle = \sum_{z,w} \alpha_{zw}|z\rangle|w\rangle$, where z is the content of T's output tape and w is the content of T's garbage tape together with an inner state. Since x is produced by T with probability at least $1 - \varepsilon$, M produces y with the probability at least $(1 - \varepsilon)(1 - \varepsilon') \geq 1 - (\varepsilon + \varepsilon') + \varepsilon\varepsilon'$, which is at least $1 - \tilde{\varepsilon}$. Since $state(M)$ is a constant and $state(T') \leq state(T) + state(M) + c$ for an appropriately chosen constant $c > 0$, we obtain $\mathrm{QDC}_{\tilde{\varepsilon}}^{p+c}(y) \leq \ell(|\phi\rangle) = m$.

(2) Let N be any 1rft. Assume that $m = \mathrm{QDC}_\varepsilon^p(y)$ with $y = N(x)$. Take $T \in \mathrm{QFST}^{\leq p(|x|)}$ and $|\phi\rangle, |\xi\rangle \in \mathcal{H}_{\mathrm{all}}$ such that $\ell(|\phi\rangle) = m$, $T(|\phi\rangle) = |\xi\rangle$, $\ell_{\mathrm{opt}}(|\xi\rangle) = |y|$, and $\|\langle\xi|y\rangle\|^2 \geq 1 - \varepsilon$. Since N is reversible, its transition between

two consecutive configurations is reversible. Hence, there exists another 1rft N^{-1} satisfying $N^{-1}(y) = x$. Thus, we first run T on $|\phi\rangle$ and then feed an output of T to N^{-1} symbol by symbol as soon as T writes symbols. This implies that $\mathrm{QDC}_\varepsilon^{p+c}(y) \leq m$ for a certain constant $c > 0$.   □

We return to Theorem 6 and give its proof.

Proof of Theorem 6. Let M be a 1qfa that computes f and let $S' = f(S)$. Since S' is $(\mathcal{M}, \mathcal{G})$-QFS-deep, for an appropriate $\mu \in \mathcal{M}$, the following holds: for any $p \in \mathcal{G}$, there exists a $q \in \mathcal{G}$ such that $p <= q$ and $\mathrm{QD}_\varepsilon^{p,q}(S'\lceil n) = \mathrm{QDC}_\varepsilon^p(S'\lceil n) - \mathrm{QDC}_\varepsilon^q(S'\lceil n) \geq \mu(n)$ for almost all n. By Lemma 7, we obtain $\mathrm{QDC}_\varepsilon^{p+c}(N(S\lceil n)) \leq \mathrm{QDC}_\varepsilon^p(S\lceil n)$ and $\mathrm{QDC}_\varepsilon^{q+c'}(S\lceil n) \leq \mathrm{QDC}_\varepsilon^q(N(S\lceil n))$. Thus, it follows that $\mathrm{QD}_\varepsilon^{p,q}(S\lceil n) = \mathrm{QDC}_\varepsilon^p(S\lceil n) - \mathrm{QDC}_\varepsilon^q(S\lceil n) \geq \mathrm{QDC}_\varepsilon^p(S\lceil n) - \mathrm{QDC}_\varepsilon^{q+c'}(N(S\lceil n)) \geq \mathrm{QDC}_\varepsilon^{p+c}(N(S\lceil n)) - \mathrm{QDC}_\varepsilon^{q+c'}(S'\lceil n) = \mathrm{QDC}_\varepsilon^{p'}(S'\lceil n) - \mathrm{QDC}_\varepsilon^{q'}(S'\lceil n) \geq \mu(n)$, where $p' = p + c$ and $q' = q + c'$.   □

4.2 Subadditivity and Monotonicity Properties

One important feature of classical Kolmogorov complexity is *subadditivity properties*. As shown in, e.g., [21], quantum Kolmogorov complexity fails to satisfy a certain form of these properties. For the proof, we will use a technical tool of b-pairing function. For each constant $b \in \mathbb{N}^+$, we define the b-*pairing function* $\langle \cdot, \cdot \rangle_b$ as follows. Given $z_1, z_2 \in \{0,1\}^*$, let $z_1 = z_{11}z_{12} \cdots z_{1t}z_{1,t+1}$ with $|z_{1i}| = b$ and $t = \lfloor |z_1|/b \rfloor$ for all $i \in [1,t]_{\mathbb{Z}}$ and $|z_{1,t+1}| < b$. Moreover, let $z_{1,t+1} = \bar{z}_1\bar{z}_2 \cdots \bar{z}_l$ with $\bar{z}_i \in \{0,1\}$ and $l = |z_{1,t+1}|$ for any $i \in [l]$. Finally, we set $\langle z_1, z_2 \rangle_b$ to be $0z_{11}0z_{12}0 \cdots 0z_{1t}10\bar{z}_10\bar{z}_20 \cdots 0\bar{z}_l11z_2$. It then follows that $|\langle z_1, z_2 \rangle_b| \leq (1 + 1/b)|z_1| + |z_2| + b + 2$.

Lemma 8. *Let $\eta \in (0,1)$ and let $\varepsilon < 1 - 1/\sqrt{2}$. Let p be any function on $\mathbb{N}$ and assume that $p(|x|) + p(|y|) \leq p(|x| + |y|)$ for any x and y. There exists a constant $c > 0$ such that $(1 + \eta)\mathrm{QDC}_\varepsilon^p(x) + \mathrm{QDC}_\varepsilon^p(y) \geq \mathrm{QDC}_{\varepsilon'}^{p+c}(xy)$ for almost all x and any y, where $\varepsilon' = \varepsilon(2 - \varepsilon)$.*

Proof. We further expand the scope of the b-pairing function into vectors in order to encode two quantum states $|\phi_1\rangle$ and $|\phi_2\rangle$ into a single quantum state. Let $|\phi_1\rangle = \sum_{z_1:|z_1|=k_1} \alpha_{z_1}|z_1\rangle$ and $|\phi_2\rangle = \sum_{z_2:|z_2|=k_2} \beta_{z_2}|z_2\rangle$. Given a number $\eta \in (0,1)$, we set $b = \lceil 2/\eta \rceil$. We then define the desired coding $|[\phi_1, \phi_2]_b\rangle$ to be $\sum_{z_1, z_2} \alpha_{z_1}\beta_{z_2}|z\rangle$, where $|z_1| = k_1$, $|z_2| = k_2$, and $z = \langle z_1, z_2 \rangle_b$. Note that $\ell(|[\phi_1, \phi_2]_b\rangle) \leq (1 + 1/b)k_1 + k_2 + b + 2$. In particular, when $k_1 \geq b(b + 2)$, we obtain $\ell(|[\phi_1, \phi_2]_b\rangle) \leq (1 + 2/b)k_1 + k_2$.

Let $k_1 = \mathrm{QDC}_\varepsilon^p(x)$ and $k_2 = \mathrm{QDC}_\varepsilon^p(y)$. Take $|\phi_1\rangle, |\phi_2\rangle \in \mathcal{H}_{\mathrm{all}}$, $T_1 \in \mathrm{QFST}^{\leq p(|x|)}$, and $T_2 \in \mathrm{QFST}^{\leq p(|y|)}$ such that $T_1(|\phi_1\rangle) = |\xi_1\rangle$, $T_2(|\phi_2\rangle) = |\xi_2\rangle$, $\ell_{\mathrm{opt}}(|\xi_1\rangle) = |x|$, $\ell_{\mathrm{opt}}(|\xi_2\rangle) = |y|$, $\||\langle \xi_1|x\rangle\|^2 \geq 1 - \varepsilon$, and $\||\langle \xi_2|y\rangle\|^2 \geq 1 - \varepsilon$. Assume that $k_1 \geq b(b + 2)$.

We want to design a new 1qft T to produce $\langle x, y \rangle$ with high probability. We set $|\phi\rangle = |[\phi_1, \phi_2]_b\rangle$. It then follows that $\ell(|\phi\rangle) \leq (1 + \frac{2}{b})|z_1| + |z_2| \leq (1 + \eta)|z_1| + |z_2| \leq (1 + \eta)k_1 + k_2$. On each input $\langle z_1, z_2 \rangle_b$ in $|\phi\rangle$, we run $T_1(z_1)$ to produce x and then run $T_2(z_2)$ to output y. Note that $state(T) \leq state(T_1) + state(T_2) + O(1) = p(|x|) + p(|y|) + O(1) \leq p(|x| + |y|) + O(1)$. In the end, T produces xy with error probability at most $(1 - \varepsilon)^2 = \varepsilon'$. Finally, we take an appropriate constant $c > 0$ so that $\mathrm{QDC}_{\varepsilon'}^{p+c}(xy) \leq \ell(|\phi\rangle) \leq (1 + \eta)k_1 + k_2$. □

The so-called *monotonicity property* holds for QFS-complexity.

Lemma 9. *Let* $\eta \in (0, 1)$*. Let* p *be any function. Given any* x*, there exists a constant* $c > 0$ *such that* $(1 + \eta)\mathrm{QDC}_{\varepsilon}^p(xy) \geq \max\{\mathrm{QDC}_{\varepsilon}^{p+c}(x), \mathrm{QDC}_{\varepsilon}^{p+c}(y)\}$ *for any* x *and* y*.*

Proof. Given a number $\eta \in (0, 1)$, we set $b = \lceil 1/\eta \rceil$. Let $m = \mathrm{QDC}_{\varepsilon}^p(xy)$ and take $T \in \mathrm{QFST}^{\leq p(|xy|)}$ and $|\phi\rangle, |\xi\rangle \in \mathcal{H}_{all}$ such that $T(|\phi\rangle) = |\xi\rangle$, $\ell_{opt}(|\xi\rangle) = |xy|$, and $\|\langle \xi | xy \rangle\|^2 \geq 1 - \varepsilon$. Let $|\phi\rangle = \sum_i \alpha_i |u_i v_i\rangle$ with $\sum_i |\alpha_i|^2 = 1$, where v_i is a string that forces $T(v_i)$ to produce a quantum state $|\xi_i\rangle$ satisfying $\ell_{opt}(|\xi_i\rangle) = |y|$.

We first show that $(1 + \eta)m \geq \mathrm{QDC}_{\varepsilon}^{p+c}(x)$ for an appropriate positive constant c. We modify T so that it reads $\langle u_i, v_i \rangle_b$, locates u_i, runs T on u_i, produces an output, and removes any other output related to v_i onto a garbage tape. We write T' for this modified T. Let $|\tilde{\phi}\rangle = \sum_i \alpha_i |\langle u_i, v_i \rangle_b\rangle$. Note that $\ell(|\tilde{\phi}\rangle) \leq (1 + \eta)m$. Since $T(|\tilde{\phi}\rangle) = |\tilde{\xi}\rangle$ and $\|\langle \tilde{\xi} | x \rangle\|^2 \geq 1 - \varepsilon$, we obtain $\mathrm{QDC}_{\varepsilon}^{p+c}(x) \leq (1 + \eta)m$, which implies the lemma. In a similar way, we can show that $(1 + \eta)m \geq \mathrm{QDC}_{\varepsilon}^{p+c}(y)$. □

One of the most distinguished features of quantum Kolmogorov complexity, in comparison with the classical one, is the complexity of t-fold quantum states. The quantum Kolmogorov complexity of multiple copies of a quantum state was already discussed in [5,9,21]. Since our setting is quite different from theirs, it is possible to prove Lemma 10.

Gács [9] introduced the notation of $\underline{H}(|\psi\rangle)$ for $-\log\langle \psi | \mu | \psi \rangle$, where μ is the quantum universal semi-density matrix, which is a density matrix but is allowed to use operators with trace less than 1. A *semimeasure* is a nonnegative real function $m(\cdot)$ satisfying $\sum_{x \in \Sigma^*} m(x) \leq 1$. Such an m is *universal* if, for any other semi-computable semimeasure ν, there exists a constant $c > 0$ such that $c\nu(x) \leq m(x)$ for all x. Moreover, let $K(w)$ denote the *prefix Kolmogorov complexity* of w.

Lemma 10. *Let* $\varepsilon \in [0, 1/2)$ *and* $n, r \in \mathbb{N}^+$*. There exists a constant* $c > 0$ *such that, for any* $|\psi\rangle \in \mathcal{H}_{2^n}$*, if* $\mathrm{QDC}_{\varepsilon}^k(|\psi\rangle) \leq r$*, then* $\underline{H}(|\psi\rangle) \leq r + K(r) + \log\frac{1}{1-\varepsilon} + c$*.*

Proof Sketch. We loosely follow [9]. Let P_r denote the projection to $\mathcal{H}_{2^r}$ and let $|\phi\rangle \in \mathcal{H}_{2^r}$. Let Π denote the projection onto the output strings and define $\Psi|\phi\rangle\langle\phi| = \Pi\rho\Pi^\dagger$, where $\rho = T|\phi\rangle\langle\phi|$. Assume that $\mathrm{QDC}_{\varepsilon}^k(|\psi\rangle) \leq r$.

Let $m(\cdot)$ denote a universal semimeasure and consider the operator $\Lambda = \sum_i m(i)2^{-i}P_i$. Note that $\sigma = m(r)2^{-r}T|\phi\rangle\langle\phi|$ satisfies $\sigma < c\mu$ for a certain constant $c > 0$. Take an eigenvalue decomposition $\rho = \sum_i p_i |i\rangle\langle i|$ with $p_1 \geq 1 - \varepsilon$ and $|\langle 1|\psi\rangle|^2 \geq 1 - 2\varepsilon$ since $|\langle\psi|\rho|\psi\rangle| \geq 1 - \varepsilon$, where p_1 is the largest eigenvalue. Since $|\phi\rangle\langle\phi| \leq P_r$, we obtain $m(r)2^{-r}|\phi\rangle\langle\phi| \leq \Lambda$.

It then follows that $\underline{H}(|\psi\rangle) = -\log\langle\psi|\mu|\psi\rangle \leq -\log\langle\psi|\sigma|\psi\rangle + c' = -\log(\sum_i m(r)2^{-r}p_i|\langle i|\psi\rangle|^2) + c' \leq -\log(m(r)2^{-r}p_1 \cdot \sum_i |\langle i|\psi\rangle|^2) + c' = -\log m(r) + r - \log(1 - \varepsilon) + c'$ for a certain constant $c' > 0$ because of $\sum_i |\langle i|\psi\rangle|^2 = 1$. Since $K(r) = -\log m(r)$, the lemma follows.   □

Proposition 11. *Let $k, n \in \mathbb{N}^+$. Let $\varepsilon \in [0, 1/2)$. There exists a constant $c > 0$ such that* $\max\{\mathrm{QDC}_\varepsilon^p(x^k) \mid x \in \Sigma^n\} \geq \frac{1}{2}\log\binom{k + 2^n - 1}{k} - c.$

Proof. For convenience, let $d = \binom{k + 2^n - 1}{k}$. Let $\varepsilon \in [0, 1/2)$. By Lemma 10, it follows that $\underline{H}(z) \leq \mathrm{QDC}_\varepsilon^k(z) + K(\mathrm{QDC}_\varepsilon^k(z)) + c \leq 2\mathrm{QDC}_\varepsilon^k(z) + c$ for a fixed constant $c > 0$ independent of z. Gács [9] proved that $\max\{\underline{H}(|\psi\rangle^{\otimes m}) \mid |\psi\rangle \in \mathcal{H}_{2^n}\} \geq \log\binom{m + 2^n - 1}{m}$. We then conclude that $\max\{\mathrm{QDC}_\varepsilon^p(x^k) \mid x \in \Sigma^n\} \geq \frac{d}{2} - \frac{c}{2}$. To derive the proposition, we rewrite $\frac{c}{2}$ as c.   □

Since $\binom{k + 2^n - 1}{k} \geq (2^n - 1)^k \geq 2^{k(n-1)}$, we obtain the following consequence.

Corollary 12. *There exists a constant $c > 0$ that satisfies the following: for any $k, n \in \mathbb{N}^+$, there exists a string $x \in \Sigma^n$ for which $\mathrm{QDC}_\varepsilon^p(x^k) \geq \frac{kn}{2} - \frac{k}{2} - c$.*

4.3 Depth/Shallowness of Infinite Sequences

We first claim that infinite random sequences are shallow. More specifically, letting $k \in \mathbb{N}^+$, an infinite sequence S is called $(\log^k, \varepsilon)$-*QFS-random* if, for any $c > 0$, there exists an index $n_c \in \mathbb{N}^+$ such that $\mathrm{QDC}_\varepsilon^{c\log n}(S\lceil n) > n - \log^k n$ holds for all $n \geq n_c$.

Lemma 13. *Let $k \in \mathbb{N}^+$. Any $(\log^k, \varepsilon)$-QFS-random infinite sequence is (polylog, log)-QFS-shallow.*

Proof. Let $k \in \mathbb{N}^+$ and let S denote any $(\log^k, \varepsilon)$-QFS-random sequence in $\{0,1\}^\omega$. Let c' denote any large positive constant. Note that, for any $c > c'$, $\mathrm{QDC}_\varepsilon^{c\log n}(S\lceil n) > n - \log^k n$ holds for almost all n. Since $\mathrm{QDC}_\varepsilon^{c'\log n}(S\lceil n) \leq n$, it follows that, for any $c > c'$, $\mathrm{QD}_\varepsilon^{c'\log n, c\log n}(S\lceil n) = \mathrm{QDC}_\varepsilon^{c'\log n}(S\lceil n) - \mathrm{QDC}_\varepsilon^{c\log n}(S\lceil n) < n - (n - \log^k n) = \log^k n$.   □

We next present the existence of infinite deep sequences. An underlying idea of the subsequent proof comes from [8,11,12] but the proof heavily lies on Lemmas 8–9 and Corollary 12.

Theorem 14. *There exists an infinite sequence that is (lin, const)-QFS-deep.*

Proof Sketch. Let $\varepsilon' = \varepsilon(2-\varepsilon)$ with $\varepsilon, \varepsilon' \in [0, 1/2)$. Let $m_0 = m_1 = 0$, $m_2 = 2$, and $m_{j+1} = m_j + 2^{m_j}$ for any $j \geq 2$. Note that $j \leq \log^{(3)} m_j$ for any $j \geq 6$. Next, we want to define S_j of length 2^{m_j} and the desired infinite sequence S is then defined as $S_0 S_1 S_2 S_3 \cdots$. Choose $k, r \in \mathbb{N}^+$ and $a \in \{0, 1\}$ satisfying $j = 2^k(1 + 2r) + a$. Let $d = 2^{m_j - 2^{k+4}}$. By Corollary 12, we take a string w_k of length $2^{2^{k+4}}$ for which $\text{QDC}_\varepsilon^k(w_k^d) \geq \frac{d|w_k|}{2} - \frac{d}{2} - c$. We set S_j to be w_k^d. Note that $|S_j| = d \cdot |w_k| = 2^{m_j}$. Let n denote any integer with $n \geq 6$ and let j denote the largest integer in $\mathbb{N}^+$ satisfying $m_{j+1} \leq n$. This j then induces (k, r, a). Note that $2^{16} \leq |w_k| \leq 2^{16j} \leq \log^{(3)} n$. We set $x = S_0 S_1 \cdots S_{j-1}$, $y = S_j$, and z is the rest of $S\lceil n$ so that $xyz = S\lceil n$. Note that $|x| = \sum_{i=0}^{j-1} 2^{m_i} = m_j$ since $j \geq 3$. By choosing (l, r') satisfying $j + 1 = 2^l(1 + 2r') + (1 - a)$, let $z = z_1 z_2$ so that $z_1 = w_l^{d'}$ and $|z_2| < |w_l|$ for an appropriate number $d' \in \mathbb{N}^+$. Note that $n = |xyz| = m_{j+1} + |z|$. Since $d'|w_l| = n - m_{j+1}$, $d' \leq \frac{n - m_{j+1}}{|w_l|}$ follows. Note that $2^{16} \leq |w_l| \leq 2^{16(j+1)}$.

It follows by Lemma 9 that $\text{QDC}_{\varepsilon'}^k(xyz) \geq \frac{1}{1+\eta} \text{QDC}_{\varepsilon'}^k(y) \geq \frac{1}{2}(\frac{d|w_k|}{2} - \frac{d}{2} - c) \geq \frac{n}{4}(1 - \frac{1}{|w_k|}) - \frac{c}{2}$ since $\eta \in (0, 1)$ and $d \leq \frac{n}{|w_k|}$. Recall from Sect. 3.1 that, for the identity machine M_{id} with $k_0 = |M_{id}|$, $\text{QDC}_0^k(u) \leq |u|$ holds for any u. Consider the following new 1qft N_k: on input $u \in \{0, 1\}^d$, output w_k^d by applying $\delta(q_0, \sigma|q_0, \sigma, w_k) = 1$, where $Q = \{q_0\}$, $\Gamma = \Theta = \{0, 1\}$, and $e = |w_k|$. Let $k_1 = |N_k|$. It then follows that $\text{QDC}_0^{k_1}(y) \leq d \leq \frac{n}{|w_k|}$. Similarly, we design N_l with $k_2 = |N_l|$ and obtain $\text{QDC}_0^{k_2}(z_1) \leq d' \leq \frac{n - m_{j+1}}{|w_l|}$. Let $k' = k_0 + k_1 + k_2$. We then obtain $\text{QDC}_\varepsilon^{k'}(x) \leq \text{QDC}_0^{k_0}(x) \leq m_j \leq \log n$ and $\text{QDC}_\varepsilon^{k'}(z_2) \leq \text{QDC}_0^{k_0}(z_2) \leq |w_l| \leq 2^{16(j+1)}$. Lemma 8 implies that $\text{QDC}_{\varepsilon'}^{k'}(z_1 z_2) \leq (1 + \eta)\text{QDC}_{\varepsilon'}^{k'}(z_1) + \text{QDC}_\varepsilon^{k'}(z_2) \leq 2d' + |w_l| < \frac{n}{128}$. Lemma 8 also implies that $\text{QDC}_\varepsilon^{k'}(xyz) \leq (1 + \eta)\text{QDC}_\varepsilon^{k'}(x) + \text{QDC}_\varepsilon^{k'}(yz) \leq (1 + \eta)[\text{QDC}_\varepsilon^{k'}(x) + \text{QDC}_\varepsilon^{k'}(y)] + \text{QDC}_\varepsilon^{k'}(z) \leq \frac{n}{64}$. Therefore, we conclude that $\text{QD}_{\varepsilon'}^{k,k'}(S\lceil n) = \text{QDC}_{\varepsilon'}^k(xyz) - \text{QDC}_\varepsilon^{k'}(xyz) \geq \frac{n}{4}(1 - \frac{1}{|w_k|}) - \frac{c}{2} - \frac{n}{64} \geq \frac{n}{8}$. Thus, S is (lin, const)-QFS-deep. □

The following statement suggests that quantum computation may be different from deterministic computation. For the statement, we define the notion of *infinitely often* $(\mu, \varepsilon, \mathcal{G})$-*quantum-finite-state deep* (or io-$(\mu, \varepsilon, \mathcal{G})$-QFS-deep) by replacing "almost all n" in the definition of $(\mu, \varepsilon, \mathcal{G})$-QFS-depth with "infinitely many n." The io-$(\mathcal{M}, \mathcal{G})$-QFS-depth is defined similarly to the $(\mathcal{M}, \mathcal{G})$-QFS-depth.

Proposition 15. *There exists an infinite sequence that is* io-(lin, const)-*QFS-deep but not* (lin, const)-*DFS-deep.*

Proof Sketch. The construction of the desired infinite sequence S is similar to the one given in the proof of Theorem 14. To be more precise, we first choose the same values m_j. Let $j \in \mathbb{N}^+$ be any number. If j is odd, then we choose a string S_j of length 2^{m_j} to satisfy $C(S_j) \geq |S_j|$, where $C(x)$ denotes the unconditional Kolmogorov complexity of x. Assume that j is even. Choose two numbers $k, r \in \mathbb{N}^+$ satisfying $j = 2^k(1 + 2r)$. Let $d = 2^{m_j - 2^{k+4}}$. Take w_k of length $2^{2^{k+4}}$ that satisfies $\mathrm{QDC}_\varepsilon^k(w_k^d) \geq \frac{d|w_k|}{2} - \frac{d}{2} - c$. We then define $S_j = w_k^d$. By a similar argument used in the proof of Theorem 14, we can show that $\mathrm{QD}_\varepsilon^{k,k'}(S\lceil m_j) \geq c'm_j$ for an appropriate constant $c' \in (0,1)$. Thus, S is io-(lin, const)-QFS-deep.

Next, we argue that S is not (lin, const)-DFS-deep. The following argument is essentially the same as [11]. Assume that S is (lin, const)-DFS-deep and consider $S\lceil m_{j+1}$ for any odd $j+1$. Note that, by considering M_{id} with $k_0 = |M_{id}|$, $\mathrm{DDC}^{k_0}(S\lceil m_{j+1}) \leq |S\lceil m_{j+1}| = m_{j+1}$. By our assumption, there is a number $k' \geq k_0$ such that $\mathrm{DDC}^{k_0}(S\lceil m_{j+1}) - \mathrm{DDC}^{k'}(S\lceil m_{j+1}) \geq cm_{j+1}$ for a certain constant $c \in (0,1)$. This implies that $\mathrm{DDC}^{k'}(S\lceil m_{j+1}) \leq m_{j+1} - cm_{j+1} \leq (1-c)m_{j+1}$. Take a 1dft M and a string w such that $M(w) = S\lceil m_{j+1}$ and $|w| = \mathrm{DDC}^{k'}(S\lceil m_{j+1})$. We then factorize w into xy so that M reads x and produces $S_0S_2\cdots S_{j-1}$. Assume that, when M reads y starting in inner state q_1 and ends in inner state q_2 by producing S_j. Note that $|x| \geq \frac{|S_0S_1\cdots S_{j-1}|}{k'} = \frac{m_j}{k'}$ since $k' \geq prod(M)$. It follows that $|y| \leq |w| - |x| \leq (1-c)m_{j+1} - \frac{m_j}{k'} \leq (1-c)2^{m_j} + (1-c-\frac{1}{k'})m_j \leq (1-c')2^{m_j}$ for another constant c' with $0 < c' < c$.

We then design a Turing machine's "program" p that works as follows. On input $w' = \langle q_1, q_2, y\rangle$, p simulates M starting in state q_1, reads y, and ends in state q_2. Clearly, p produces S_j. Note that $|w| \leq 2(|q_1| + |q_2|) + |y| \leq |y| + 2k' < (1-c'')2^{m_j}$ for an appropriate constant c'' with $0 < c'' < 1$ since $|q_1|, |q_2| \leq k'$. This implies that $C(S_j) \leq |w'| \leq (1-c'')2^{m_j} = (1-c'')|S_j|$. This contradicts $C(S_j) \geq |S_j|$. $\qquad\square$

5   A Brief Closing Discussion

Throughout this work, we have intended to expand the scope of Bennett's notion of logical depth and shallowness of binary static data to a realm of quantum computing. In particular, this exposition has introduced the notions of quantum finite-state (QFS) depth and shallowness, which are based on quantum finite-state transducers, aiming at the comprehensive study on compression/decompression procedures of streaming data coming in incessantly through a communication channel. It should be remarked that our QFS-depth notion does not require the existence of universal machine. This may open a door to an introduction of a similar depth notion to other models of computations despite the lack of universality.

Nonetheless, we are still at an initial stage of the full fledged research on the foundations and applications of practical variants of logical depth and shallowness on the medium of quantum computing. Our future goal is to cultivate the theoretical foundations and find various practical applications in different fields of science and engineering.

References

1. Antunes, L., Fortnow, L., van Melkebeek, Vinodchandran, N.: Computational depth: concept and applications. Theor. Comput. Sci. **354**, 391–404 (2006)
2. Benioff, P.: The computer as a physical system: a microscopic quantum mechanical Hamiltonian model of computers represented by Turing machines. J. Statist. Phys. **22**, 563–591 (1980)
3. Bennett, C.H.: The Universal Turing Machine, A Half-Century Survey, chap. Logical Depth and Pysical Complexity, pp. 227–257. Oxford University Press, Oxford (1988)
4. Bernstein, E., Vazirani, U.: Quantum complexity theory. SIAM J. Comput. **26**, 1411–1473 (1997)
5. Berthiaume, A., van Dam, W., Laplante, S.: Quantum Kolmogorov complexity. J. Comput. System Sci. **63**, 201–221 (2001)
6. Deutsch, D.: Quantum Theory, the Church-Turing Principle, and the Universal Quantum Computer. Proc. R. Soc. Lond. Ser. A **400**, 97–117 (1985)
7. Deutsch, D.: Quantum computational networks. Proc. R. Soc. Lond. Ser A **425**, 73–90 (1989)
8. Doty, D., Moser, P.: Feasible depth. In: Cooper, S.B., Löwe, B., Sorbi, A. (eds.) CiE 2007. LNCS, vol. 4497, pp. 228–237. Springer, Heidelberg (2007). https://doi.org/10.1007/978-3-540-73001-9_24
9. Gács, P.: Quantum algorithmic entropy. J. Phys. A Math. Gen. **34**, 6859–6880 (2001)
10. Huffman, D.A.: Canonical forms for information-lossless finite-state logical machines. IRE Trans. Circ. Theory CT **6**, 41–59 (1959)
11. Jordon, L., Moser, P.: On the difference between finite-state and pushdown depth. In: Chatzigeorgiou, A., et al. (eds.) SOFSEM 2020. LNCS, vol. 12011, pp. 187–198. Springer, Cham (2020). https://doi.org/10.1007/978-3-030-38919-2_16
12. Jordon, L., Moser, P.: Pushdown and Lempel-Ziv depth (2020), Manuscript. arXiv:2009.64821v1
13. Juedes, D.W., Lathrop, J.I., Lutz, J.H.: Computational depth and reducibility. Theor. Comput. Sci. **132**, 37–70 (1994)
14. Kitaev, A.Y., Shen, A.H., Vyalyi, M.N.: Classical and Quantum Computation, Graduate Studies in Mathematics. Americal Mathematical Society, Providence (2002)
15. Kolmogorov, A.N.: Three approaches to the quantitative definition of information. Probl. Inf. Transmis. **1**, 1–7 (1965)
16. Li, M., Vitányi, P.M.: An Introduction to Kolmogorov Complexity and Its Applications, 3rd edn. Springer-Verlag, Berlin (2008). https://doi.org/10.1007/978-0-387-49820-1
17. Moser, P.: Polylog depth, highness and lowness for E. Inf. Comput. **271**, 104483 (2020)
18. Nielsen, M.A., Chuang, I.L.: Quantum Computation and Quantum Information. Cambridge University Press, Cambridge (2000)
19. Nishimura, H., Yamakami, T.: Polynomial time quantum computation with advice. Inf. Process. Lett. **90**, 195–204 (2004)
20. Shannon, C.E.: A mathematical theory of communication. Bell Syst. Tech. J. **27**(379–423), 623–656 (1948)
21. Vitányi, P.M.B.: Quantum Kolmogorov complexity based on classical descriptions. IEEE Trans. Inf. Theor. **47**(6), 2464–2479 (2001)

22. Yamakami, T.: A foundation of programming a multi-tape quantum turing machine. In: Kutyłowski, M., Pacholski, L., Wierzbicki, T. (eds.) MFCS 1999. LNCS, vol. 1672, pp. 430–441. Springer, Heidelberg (1999). https://doi.org/10.1007/3-540-48340-3_39

23. Yamakami, T.: One-way reversible and quantum finite-state automata with advice. Inf. Comput. **239**, 122–148 (2014)

24. Yamakami, T.: Relativizations of nonuniform quantum finite automata families. In: McQuillan, I., Seki, S. (eds.) UCNC 2019. LNCS, vol. 11493, pp. 257–271. Springer, Cham (2019). https://doi.org/10.1007/978-3-030-19311-9_20

25. Yamakami, T.: A schematic definition of quantum polynomial time computability. J. Symb. Logic **85**, 1546–1587 (2020)

26. Yamakami, T.: Nonuniform families of polynomial-size quantum finite-state automata and quantum logarithmic-space computation with polynomial-size advice. Inf. Comput. (in press, 2021), A preliminary version appeared in the Proceedings of LATA 2019, LNCS, vol. 11417, pp. 134–145 (2019). A complete and corrected version. arXiv:1907.02916

27. Yao, A.C.: Quantum circuit complexity. In: Proceedings of the 34th Annual IEEE Symposium on Foundations of Computer Science (FOCS 1993), pp. 80–91 (1993)

28. Ziv, J., Lempel, A.: Compression of individual sequences via variable-rate coding. IEEE Trans. Inf. Theor. **24**, 530–536 (1978)

Author Index

Caballero, David 1

Dale, Matthew 19

Fletcher, Willem 35

Gomez, Timothy 1

Hader, Daniel 116

Isuzugawa, Raimu 51
Itoh, Toshiya 149

Khadiev, Kamil 68
Khadieva, Aliya 84
Klinge, Titus H. 35
Kravchenko, Dmitry 68
Kutrib, Martin 101

Lathrop, James I. 35

Miyahara, Daiki 51
Mizuki, Takaaki 51

Nye, Dawn A. 35

O'Keefe, Simon 19

Patitz, Matthew J. 116

Rayman, Matthew 35
Roychowdhury, Jaijeet 131
Ruangwises, Suthee 149

Schweller, Robert 1
Sebald, Angelika 19
Stepney, Susan 19, 164
Summers, Scott M. 116

Trefzer, Martin A. 19

Wendlandt, Matthias 101
Wylie, Tim 1

Yakaryılmaz, Abuzer 84
Yamakami, Tomoyuki 177

Printed in the United States
by Baker & Taylor Publisher Services

Printed in the United States
by Baker & Taylor Publisher Services